The Institutionalisation of Disaster Risk Reduction

The past three decades have seen a global shift in disaster management from an event driven response to a 'could-be' risk management approach. Disaster risk reduction (DRR) has become entrenched as a dominant paradigm within the field of disaster management.

More than a decade after adopting DRR legislation in South Africa there remains a dearth of evidence that this has translated into substantive action. This book examines the institutionalisation of DRR in South Africa, conceived of as a political economy of knowledge production. Using a critical theory approach, the book does not consider why DRR is failing but instead asks 'why DRR?' As such, it explores possibilities beyond DRR's narrow optic and offers new insights into disaster management through the lens of South Africa.

This is valuable reading for graduate students and academics working in disaster studies, geography, public policy and development/post-development studies, as well as policy makers.

Gideon van Riet is Senior Lecturer in Political Studies at North-West University's Potchefstroom Campus in South Africa.

Routledge Studies in Hazards, Disaster Risk and Climate Change

Series Editor: Ilan Kelman Reader in Risk, Resilience and Global Health at the Institute for Risk and Disaster Reduction (IRDR) and the Institute for Global Health (IGH), University College London (UCL)

This series provides a forum for original and vibrant research. It offers contributions from each of these communities as well as innovative titles that examine the links between hazards, disasters and climate change, to bring these schools of thought closer together. This series promotes interdisciplinary scholarly work that is empirically and theoretically informed, with titles reflecting the wealth of research being undertaken in these diverse and exciting fields.

Published

Cultures and Disasters
Understanding Cultural Framings in Disaster Risk Reduction
Edited by Fred Krüger, Greg Bankoff, Terry Cannon, Benedikt Orlowski and E. Lisa F. Schipper

Recovery from Disasters
Ian Davis and David Alexander

Men, Masculinities and Disaster
Edited by Elaine Enarson and Bob Pease

Unravelling the Fukushima Disaster
Edited by Mitsuo Yamakawa and Daisaku Yamamoto

Rebuilding Fukushima
Edited by Mitsuo Yamakawa and Daisaku Yamamoto

Climate Hazard Crises in Asian Societies and Environments
Edited by Troy Sternberg

The Institutionalisation of Disaster Risk Reduction
South Africa and Neoliberal Governmentality
Gideon van Riet

The Institutionalisation of Disaster Risk Reduction

South Africa and Neoliberal Governmentality

Gideon van Riet

Routledge
Taylor & Francis Group

LONDON AND NEW YORK

First published 2017
by Routledge
2 Park Square, Milton Park, Abingdon, Oxon OX14 4RN

and by Routledge
52 Vanderbilt Avenue, New York, NY 10017

First issued in paperback 2020

Routledge is an imprint of the Taylor & Francis Group, an informa business

© 2017 Gideon van Riet

British Library Cataloguing in Publication Data
A catalogue record for this book is available from the British Library

Library of Congress Cataloging in Publication Data
A catalog record for this book has been requested

ISBN 13: 978-0-367-67051-1 (pbk)
ISBN 13: 978-1-138-20677-9 (hbk)

Typeset in Times New Roman
by Apex CoVantage, LLC

Contents

Illustrations

Acknowledgements

I would like to thank Cecile Schultz, my loving fiancée and companion. I would also like to thank Dawie van Vuuren from MetroGIS for producing the map of South Africa free of charge.

Abbreviations

AIDS	Acquired Immune Deficiency Syndrome
ANC	African National Congress
BEE	black economic empowerment
CBDRA	community-based disaster risk assessment
CBDRM	community-based disaster risk management
CoCT	City of Cape Town
CoT	City of Tshwane
CODESA	Convention for a Democratic South Africa
CSIR	Council for Scientific and Industrial Research
CSR	corporate social responsibility
DESNOS	disorders of extreme stress not otherwise specified
DMC	Disaster Management Centre
DiMP	Disaster Mitigation for sustainable livelihoods Programme
DoHE	Department of Higher Education
DMA	Disaster Management Act 57 of 2002
DMCs	disaster management centres
DRA	disaster risk assessment
DRAaM	disaster risk assessment and management (a neologism)
DRM	disaster risk management
DRMP	disaster risk management plan
DRR	disaster risk reduction
EGS	environment and geographical sciences
EDM	Eden District Municipality
GEAR	Growth Employment and Redistribution
GIS	geographical information system
HFA	Hyogo Framework for Action
HIV	Human Immunodeficiency Virus
HSRC	Human Sciences Research Council
ICT	information communication technology
IDP	integrated development plan

IDNDR	International Decade for Natural Disaster Reduction
IPCC	International Panel on Climate Change
KPA	key performance area
MDGs	Millennium Development Goals
NDMF	National Disaster Management Framework
NGO	non-governmental organisation
NPAI	New Public Management Initiative
NPM	new public management
NQF	National Qualifications Framework
PAR	Pressure and Release Model
PTSD	post-traumatic stress disorder
RADAR	risk and development review
RDP	Reconstruction and Development Programme
SLF	Sustainable Livelihoods Framework
SAP	Structural Adjustment Programme
UDW	University of Durban-Westville
UNDP	United Nations Development Programme
UNISA	University of South Africa
UNISDR	United Nations International Strategy for Disaster Reduction
UWC	University of the Western Cape
WCED	World Commission on the Environment and Development
WSSD	World Summit on Sustainable Development

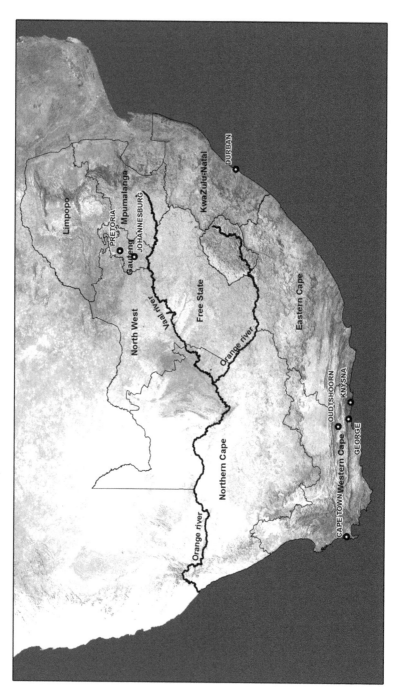

Map 1 South Africa

Introduction

South Africa is often praised for her progressive and comprehensive disaster management legislation. The legislation in question adheres to the prevailing principles of disaster risk reduction (DRR), a preeminent approach to disaster management emerging in the last quarter of the 20th century, which has been entrenched in various international policy conventions (see Chapter 1). As such, the country officially follows a developmental and risk management approach to disaster management by focusing on aspects such as prevention, mitigation and preparedness, as opposed to merely disaster response. In addition to the Disaster Management Act 57 of 2002 (hereafter the DMA), officially passed in January 2003, a comprehensive policy framework offering more detailed guidance to practitioners followed in 2005 (South Africa, 2005). These policy documents outline required and recommended institutions for integrating DRR into development. However, it seems that the result has been less than satisfactory. Existing evidence suggests that implementation of this legislation has been limited, at best (Botha et al., 2011).

This book explores the institutionalisation of this relatively recent approach to disaster management in South Africa. For the purposes of the current analysis the question of why DRR is failing is largely avoided. Rather, the underlying question implied throughout is, 'why DRR?' This avoids the inevitable answer of the former question, attributing failure to some or other articulation of 'poor policy implementation' or a 'lack of political will' (cf. Botha et al., 2011; Wisner et al., 2011). Instead, by working within the recognised tradition of critical theory, DRR is treated as inherently suspect. Such a strategy opens up possibilities beyond DRR's narrow optic and its celebrated policy documents, often treated as universal dogma across space and time, by practitioners and academics alike.

Disaster risk reduction in South Africa emerged in the 1990s amid a changing national context. Democratisation meant that old modes of governing the state and society needed to be replaced, while this transition afforded an opportunity to end the country's relative isolation. Greater incorporation followed, into what may be described as a global system largely informed by neoliberalism. Neoliberalism has been defined as a set of political assumptions based on the belief that the only legitimate purpose of the state is to safeguard the individual's commercial and private property rights (Harvey, 2005:2). This is best achieved where exchange of goods and services are organised on free-market principles in accordance with the writings of classical economists such as Adam Smith (2005). Adopting such free-market principles is meant to provide a context that is most conducive to individuals realising their natural creative potential and entrepreneurial spirit. It will therefore increase individual liberty and well-being and affect a more efficient allocation of resources (see Friedman, 2006).

Neoliberalism came to prominence amid the decline of the welfare state, which had, according to two prominent leaders in the form of Margaret Thatcher and Ronald Reagan, made people lazy and not the free-enterprising creative individuals described earlier. Neoliberalism has significant global dimensions, maintained through the Bretton Woods institutions. The most notorious of these is arguably the role played by the World Bank and the International Monetary Fund (IMF), in imposing a particular set of neo-liberal economic policies – the so-called Washington Consensus – on vast parts of the Global South. Especially the African countries hardest hit by the oil price increases of 1973 and 1978 were subjected to the Washington Consensus. The Structural Adjustment Programmes (SAPs) instituted by the IMF and the World Bank provided countries with balance-of-payment support and funding for critical development projects, tied to specific conditions. These conditions included privatisation, minimal state intervention in the economy, significant reduction of trade barriers, liberal democratic reforms and even changes in what was produced within the country. Many authors have, however, commented on the general failure of these initiatives, as countries subjected to SAPs become more indebted, while indicators such as infant mortality, literacy levels and malnutrition worsened, and the poverty rates of and economic inequality within the countries that adopted such policies (and between North and South) increased (see Harvey, 2005). Since the end of the Cold War, neoliberal economics became even more influential. The 'triumph' of liberal democracy over Communism, according to Fukuyama (1989), marked the 'End of History'. This sentiment offered much justification for neoliberal economics' purported status as 'the way'. The transformation towards a post-apartheid South Africa has been informed significantly by neoliberal conventional wisdom. This is evident in the emphasis on macroeconomic stability, specifically keeping inflation within strict parameters as the principle objective of economic policy. Furthermore, the extensive outsourcing of public services

(Rogerson, 2010) is another obvious manifestation of neoliberalism's pertinence to the current study.

Disaster risk assessment and management in South Africa, especially at local government level, has largely been commodified and outsourced. The extent of this approach to disaster management suggests that the local manifestation of the global DRR epistemic community is best viewed as an industry. The state's reliance on consultants implies a form of state building in the 'knowledge economy', whereby state building is conducted through the use of non-state actors, in the form of so-called experts. It also implies a particular political economy of knowledge production that includes academics, consultants, state officials, ordinary citizens and hybrids of the above. All of these actors have interests in DRR or its operationalisation, disaster risk management (DRM). In this regard we may for example consider the profit motive, service to society and exposure to material dangers. For some, certain objectives are more important than they are for others. These objectives and how they are sanctioned are also part of, and continue to shape, the discursive context in which DRM is continually produced and reproduced.

To conceive of disaster management in the mode of DRR is to frame the problem of disaster in the idiom of risk assessment and management (hereafter *disaster risk assessment and management* or *DRAaM*). Disaster risk assessment and management shares various characteristics with other discourses, institutionalised since the end of the Second World War. Such discourses include: security, gender, public health and development, including numerous sub-fields such as development economics, food policy, indigenous knowledge, (good) governance; (cf. Berridge, 2007; Doornbos, 2003; Escobar, 1988; Laurie et al., 2005) and many more examples. There appear to be common trends pertinent to DRAaM. Once an initial idea is conceived, many factors facilitate its institutionalisation. This includes professionalisation, *scientisation* (subjecting the idea to the practices of science) and commodification, while the development of important platforms, such as journals and conferences, may play a solidifying role. Epistemic communities mobilise around a 'cutting-edge' idea and gather within or across disciplines, seemingly serving a common professional agenda. This often implies both a narrowing of established disciplines and what is construed as 'boundary work', cutting across disciplines. It almost always entails a more 'practical' or applied focus, which may be both all-embracing and totalising and superficial and vague.

New experts typically find employment in think-tanks, in academia or as consultants. The latter are increasingly pervasive. Contemporary modern society holds them in high regard, and hence they are a quintessential feature of the contemporary knowledge-based neoliberal epoch. Experts draw on the codified theoretical knowledge Daniel Bell (1973) refers to in explaining the then-impending post-industrial society (in recent years more commonly known as the knowledge economy). These experts are privy to a particular code, shared by their particular epistemic community. The code may be viewed as a preconceived conceptual framework used to interpret any real-life problems authorised by a paying client. Such perspectives have been criticised exactly for degenerating into all-encompassing discourses. For example, in security studies the concept *securitisation* was coined by the so-called Copenhagen School, critical of the fact that increasingly diverse matters are framed as 'security' concerns (Buzan et al., 1998). The result is an interpretive slippage as a broadly accepted denotation is gradually eroded into a state of relative emptiness. Especially in the case of a multidisciplinary field, these 'experts' better resemble generalists, who possess little understanding of traditional disciplinary knowledge. This often implies limited theoretical depth (and therefore interpretive acumen) in the knowledge they produce (Laurie et al., 2005:477).

The shared code referred to earlier, it might be argued, drawing on Adorno and Horkheimer's (1969) *Dialectic of Enlightenment*, can be central in shaping the identities of practitioners in relation to their expert field, in what Laurie et al. (2005:479) refer to as 'governable' – and one could add 'intellectual' – spaces. These governed intellectual spaces are fields of expert knowledge characterised by particular discursive and practical styles, which proponents believe help humans address the prevalent threats produced by everyday socio-environmental processes. Adorno and Horkheimer refer to these secular processes as *nature*. In this effort to dominate nature, actors often fall prey to instrumental reason, which may be defined as reason in the service of the aforementioned preconceived interpretations of the world. In other words, experts tend to ask 'how' to apply a particular conceptual code as opposed to 'why'. Often the result is contextually inappropriate intervention.

Though the use of *Dialectic of Enlightenment* must be qualified significantly, there are particular reasons for this choice. The text is perhaps dated if one considers the particular context of national socialism in Germany during which it was written. However, this text offers a general narrative which fits the subject matter of this book and the type of analysis provided. The storyline the reader can expect is one in which rationality has produced a particular concept of emancipation. Unfortunately, this idea has been subjected to instrumental rationality, resulting in a state of 'barbarism'. Adorno and Horkheimer's rather tragic tale offers a unique means of emphasising the often depraved consequences of this particular instrumental reason in the South African context. This narrative therefore, in addition to offering a generally apt description of the subject matter, provides an interesting way of interjecting pathos into the analysis where the data demands it. To reiterate, the general framework provided by such an interpretation of the *Dialectic of Enlightenment* is supplemented with analytical constructs that allow for more precise explanation. The most important of these theoretical tools is the Foucauldian notion of power.

Governed intellectual spaces, such as DRAaM, are the locales for intended and unintended gatherings of particular words, meanings and subjectivities. They are, as explained, the terrain of expert knowledge and facilitate a distinct sense of self. Thus, following Foucault (1982), it can be argued that institutionalised or institutionalising discourses are intimately linked to power, and through these intertwined knowledge–power dynamics they facilitate specific modes of governmentality within each of these fields. 'Governmentality' is a broader term than 'government' or more conventional definitions of governance. It is a function of various intended and unintended actions, reflections and interactions, where power is exercised, influencing the day-to-day actions of individuals, groups and institutions both directly and indirectly. Governmentality is very much a function of the structuring effects of the combined forces of discourse and power or knowledge-power, for example in constituting 'subjects' such as DRAaM practitioners. The word 'subject' therefore requires inverted commas, as there can be no truly independent agency. Based upon these first pages of introductory discussion, the implications of the foregoing as a statement of intent are summarised in what follows and thereafter developed further through discussions on matters of ontology, epistemology, methodology, analytical style and structure.

Purpose and method

This book offers a critical investigation into DRAaM's institutionalisation in South Africa. My argument is that that industry is essentially a manifestation of the aforementioned, political economy of knowledge production set in a broad context of neoliberal governmentality. Neoliberalism has been a critical discourse through which power is exercised across the globe and has similarly informed many other governing discourses. Disaster risk assessment and management in South Africa has been institutionalised in this context, where markets for expert forms of knowledge are an important feature. As such DRAaM has manifested as a form of expert knowledge around which a local industry of consultancy, research and education has developed associated with various markers and performances of professionalism. By framing the disaster problem in the idiom of risk assessment and management, the result has been often questionable knowledge production and education characterised by instrumental reason. As a result of this instrumentalism, the initial objective of service to humanity has been largely lost, and DRAaM has regressed into multiple dystopias for both its officially intended beneficiaries and DRAaM practitioners.

To avoid misinterpretation it must be emphasised that my analysis should not in any way be construed as 'disaster *denialism*'. Instead of denying the problem of disaster or the sentiments from which the initial idea of disaster reduction emerged, the argument is simply that this industry, created to alleviate very real dangers, has taken on a form that is not very beneficial, if not at times detrimental, to the initial cause. The book does not offer the alternative recipe for emancipation a critic might demand. That would be impossible, as such prescription cannot possibly contend with the uniqueness of each context. The book holds certain arguments at the aggregate or national level of analysis, but it recognises that meaningful interventions at lower levels can be tailored only through consistent and open-ended dialogue, with ordinary citizens contributing their particular interpretations of a context of which they are perhaps most knowledgeable.

The book draws on a mixture of qualitative methods. An extensive literature review was conducted including local and international policy documents, university curricula, academic research in DRAaM and consultants' reports. In addition to the literature on DRAaM the review included literature on numerous related topics such as housing, migration, public health, food security and poverty and inequality. The objective here was to form a better picture of the socio-economic context to which DRAaM is meant to contribute. The analysis is also informed by primary data in the form of semi-structured interviews with relevant actors, such as consultants and academics, and some participant observation. Regarding the latter I attended two regional conferences. The primary data offers an indispensable set of insider perspectives.

Instrumental reason and neoliberal governmentality: towards an analytical framework

The book adopts a similar narrative to Adorno and Horkheimer's (1969), of enlightenment leading to myth and barbarism through instrumental reason. This very general explanation is however supplemented most notably with Foucault's conception of structuring power and the associated concept of governmentality, to allow greater precision and to alleviate what are perhaps problematic foundations of Adorno and Horkheimer's text, its implicit ahistorical Freudian determinism. In particular, the Foucauldian notion of structuring power has superior acumen for dealing with the delicate and dynamic complexities of subject construction.

I believe the sentiment of Adorno and Horkheimer's text is relevant, first because of how DRAaM implies objectifying and dominating nature in the name of (physical) self-preservation. It is argued that many of the motives and consequences of this type of reasoning, about which Adorno and Horkheimer theorise, apply to DRAaM in South Africa. Still significant qualification is required with regard to the application of *Dialectic of Enlightenment* in a much different time and space compared to the context Adorno and Horkheimer wanted to account for. The theme of emancipation through institutionalised reason, however, remains. Adorno and Horkheimer argue that reason leads to myth, through totalising discourse and 'childish science', which saps creativity from thought and reverts to instrumental reason in an effort to maintain (a sense of) self-preservation. The difference is first that DRAaM, as a manifestation of post-apartheid emancipatory praxes, is located in a context of neoliberal governmentality and not in the national socialism and totalitarian Marxism that characterised different parts of Europe in the 1930s and 1940s.

Second, the underlying ontology on which this analysis is premised is different to that of Adorno and Horkheimer's analysis. Foucault's perspective on power as a structuring element in society is employed in order to explain how subjects such as DRAaM practitioners and therefore the industry in general are constituted. One might for example ask exactly how discourses such as DRAaM lead to the quest for self-preservation via instrumental reason. The *Dialectic of Enlightenment*'s Freudian underpinning implies that certain types of behaviour, such as those based on a need for self-preservation, are inherent across space and time. This potentially leads to ahistorical and superficial analysis and even the very preordained instrumentality of which Adorno and Horkheimer were so critical. Thus, I want to argue that this need is both constituted by and a vehicle for power. There is more to this particular picture or moment in time and space than can be revealed by the concepts of self, id, ego and superego.

Therefore, while following the *Dialectic of Enlightenment*'s basic logic, Foucault's notion of power is interjected to add more nuance and detail to the *Dialectic of Enlightenment*. To this end, various micro-discourses constitutive of DRAaM, such as science, expert knowledge, community-based disaster risk management, geographical information systems (GIS), and legislative compliance are scrutinised for their influence on processes of subject formation. Foucault's conception of power is useful in navigating the numerous shades between often overlapping categories such as 'wittingly' and 'unwittingly', when applied to an industry, which arguably still includes noble intent. When action is viewed through a Foucauldian lens 'intent' becomes a rather problematic term, not to be conceived of in absolute terms. There is no subject per se, since agency is never independent of its more or less fleeting milieu, structured by power. The book describes how the domination of certain groups over others is no exception, due to the ignorance produced by certain prevalent meta-narratives (or macro-discourses) such as risk assessment and management and neoliberalism, in conjunction with the aforementioned micro-discourses. The logical conclusion of this adaptation to the *Dialectic of Enlightenment* is that the text is relieved of its ahistorical and relatively insubstantial (for our purposes at least) Freudian determinism. In the case of DRAaM, enlightenment definitely leads to myth. However, the process unfolds under specific contextually determined circumstances through key actors constituting discourses and practices. This detail matters very much, though it can only be revealed to any meaningful degree by adding more theoretical tools.

A substantial social scientific body of literature exists on the concept of risk and associated epistemologies, which seemingly constitute another contemporary global metanarrative. The idiom of risk assessment and management is applied in industries to navigate issues ranging from insurance and investments to environmental management, nursing, business ethics and occupational health and safety. Consequently, the list of analytical materials discussed in this chapter cannot be complete without explicitly engaging with how risk management more generally manifests as a prominent feature of late-modern society. This brief section outlines the theoretical approach taken in critically analysing how the concept 'risk' is applied in DRAaM.

In order to yield a meaningful critique of the South African DRAaM industry, the concept of 'risk' cannot be taken for granted. Such a positivistic approach would eliminate many potential avenues for fruitful critical analysis. In order to open up the potential for critique from outside the idiom of risk assessment and management while not falling prey to the disaster denialism mentioned previously, the work of cultural anthropologist Mary Douglas (1991) is useful. Douglas distinguishes between *risk* and *danger*. She argues that in the West risk has historically been discussed as synonymous with danger. This conflation of terms is unfortunate, as risk is also associated with probability theory and related analytical technologies. In order to not lose sight of the very real hardships many must endure because of disaster, while allowing for critique of the practice of risk assessment and management, a distinction is required between danger as a reality for many, and risk, which is inextricably linked to the idiom of risk assessment and management. Risk, accordingly, is associated with the measurement of danger, in an attempt to dominate nature. However, although these conditions of danger are very real, they are difficult if not impossible to objectively measure. This distinction therefore allows a researcher to stand outside of the idiom of risk assessment and management in order to critique it. The loaded term 'risk' and its associated analytical practices are necessarily subject to further scrutiny. To further develop a critique of the concept of risk, a systematic definition of a concept related to the framing of a problem in the idiom of risk assessment and management is required. For this the neologism *risk framing* was chosen. Risk framing refers to the framing of a matter as one for risk assessment and management and can be explained by way of a three-tiered definition.

It is first the *de-politicising* of material dangers by framing an issue in the idiom of risk assessment and management. Hence, it is a type of reification. The fundamental socio-political causes of danger are denied, obscured or downplayed and reframed as a technical management issue. This may be contrasted with *securitisation*. Buzan et al. (1998), in coining securitisation theory, argue that by framing an issue as a matter of security or rather *for* security, that speech act, if bought into by relevant stakeholders, removes the issue from the realm of the 'normal politics' of deliberation and places it in the realm of 'emergency politics'.

For those in a position to access agenda-setting power, open discussion and negotiation, a defining characteristic of normal politics, the matter is no longer an option. Whereas securitisation implies that an issue is elevated to an immediate and urgent political problem in the realm of emergency politics, risk framing, despite rhetoric to the contrary, implies de-prioritisation. Real material dangers are deemed essentially less urgent as they become long-term management concerns. Framing an issue as a concern for risk assessment and management also takes it out of the realm of normal politics, though in an opposite direction when compared to securitisation. The matter is still not open for debate and discussion as it has already been delegated to those occupied with a particular mode of technocracy and bureaucratic management.

From a Foucauldian perspective, however, if power is omnipresent, then no social context can ever be apolitical. The very act of de-politicising an issue, in reshuffling power dynamics and affirming particular macro-discourses such as DRAaM, is also to re-politicise the matter. In the second place, it can therefore be argued that risk framing implies re-politicisation, which may manifest in numerous ways. To narrow the current discussion down for the purposes of this book Bankoff's (2001) analysis of vulnerability as a Western discourse used to 'denigrate large parts of the world' is instructive. Bankoff argues that discourse associated with DRR is the latest manifestation of a trend previously displayed as discourses of tropicality and later as underdevelopment. Vast parts of the world are labelled as vulnerable or at risk when compared to an implicit norm based on a Western frame of reference. Bankoff argues that such labelling practices are decidedly political, as the distinction between an 'us' and a 'them' is reiterated as an asymmetrical relationship. Specifically, this type of labelling by those with access to specialist codes individualises vast proportions of the globe by highlighting the deviance and implied inferiority of those who are

subjected to these labelling practices. Such labelling practices have historically emerged in the West and have largely been governed from the West.
By drawing on some of the previous discussions one might argue that these
practices are exercised from a position of power due to experts' privileged
access to a particular type of codified theoretical knowledge. These discourses of vulnerability or being 'at risk' create the potential for simplistic
reason in the service of a risk assessment and management-related code
experts are privy to.

The third and final tier of our definition pertains to how the implementation of the idiom of risk assessment and management might feed back into
the materiality of danger. As a manifestation of instrumental reason and a
predefined and very general explanation of reality, interventions may serve
universalising and externally imposed conceptions of local realities, for
example, based upon international conventions and/or national-level policy
documents. Where this is the case, the effects of interventions on the materiality of danger are likely to be minimal at best and detrimental at worst.

Chapter outline

The remainder of the book is structured as six chapters followed by a concluding chapter. Chapters 1 and 6 are independent units. Chapters 3 and 4
and 5 and 6 form two sets of two chapters each. The first of these sets is
more concerned with matters of danger and disaster and how these relate
to the broader South African political economy. The second set, Chapters 5
and 6, focus on DRAaM's institutionalisation in general and, in the case of
Chapter 6, DRAaM knowledge production in particular.

Chapter 1 provides historical context in the form of a genealogy of
DRAaM. In particular, the chapter traces the institutionalisation of DRAaM
globally along two broad bodies of literature. The first body of literature is
that of development. The second is associated with the global meta-narrative
of risk assessment and management. It is argued that the initial emancipatory objectives of disaster reduction have been corrupted by the overly
bureaucratising inklings of instrumental rationality.

Chapters 2 and 3 analyse the South African context in which 'disaster
risk' is said to manifest. The DRAaM discourse is first approached by
drawing on DRAaM vernacular to produce a hypothetical risk profile of
the country. However, such an understanding of disaster and intertwined
matters of concern is unsatisfactory and, as such, these various interlinked
problematics require an alternative conceptualisation to emphasise the
urgency and complexity of prevalent dangers and associated, often overlapping, socio-political circumstances many South Africans face. Chapter
3 provides such an alternative by defining the everyday as situated within

structural violence. This conceptualisation offers a position from which the South African DRAaM industry can be critiqued in Chapters 4, 5 and 6.

Chapter 4 provides an institutional analysis of DRAaM in South Africa, including both an analysis of individuals and institutions active in DRAaM and what it means to be engaged in DRAaM in South Africa. The chapter therefore maps the industry by analysing DRAaM institutions, where 'institution' is viewed both as a noun and a verb. In the process, pertinent features of state building and higher education in a context of neoliberal governmentality are highlighted, with particular emphasis on the interwoven relationship between practice and higher education.

The focus of Chapter 5 is on the types of knowledge produced in the South African DRAaM industry. The discussion is divided into academic knowledge and consultancy-based knowledge and further divided along the most significant actor and industry-constituting micro-discourses. This knowledge is often produced through instrumental reason, where policy documents and a very simplistic, vague and all-encompassing expert code inform instrumental logics through governing discourses such as legislative compliance, science, multidisciplinarity, community-based DRM and GIS. The result is that the knowledge produced is often of poor quality and therefore not particularly useful.

Chapter 6 considers the consequences of DRAaM's institutionalisation in South Africa. It is argued that DRAaM has regressed into a set of multiple dystopias for both practitioner and intended beneficiary. While structural violence endures for many South Africans, experts are also more or less troubled by the relative impotence of their labour. Here the concept *anomie*, in particular the conception of the term Robert Merton (1938) devised to explain deviant behaviour, is preferred to the Marxist notion of alienation to explain this disenchantment. *Anomie* following from a disjuncture between industry-specific objectives and legitimate means by which to achieve these is both more precise and less fixated on an essential 'human nature' implied in notions such as 'alienated from yourself'. Still, by drawing on Merton in this way, it is still possible to capture a similar emotion. The analysis largely attributes the disjuncture between means and ends to the competitive industry and the imperative to undercut competitors with the obvious result that consultancy projects are rushed, though even if this were not the case, the extreme skills set implied by the notion of a 'DRM practitioner' would have remained a difficult if not an impossible achievement.

The concluding chapter draws together insights from the preceding chapters. It offers concluding remarks on DRAaM in South Africa but also more generally with regard to South African politics and the politics of north–south relations as they relate to knowledge production. It appears that attempts at preserving the self and its professional interests are futile.

Instead, the slightest possibility of progressive change lies only in the opposite: (metaphorical) self-destruction. To build the state, we must destroy the status quo and, by implication, our power-wielding self and all of its darlings. The professional identities emerging from institutionalisation such as that depicted in this book and the means by which they emerge must be challenged and rejected where they are inappropriate or otherwise problematic, as there can be no universal and timeless professional identities.

References

Adorno, T. and Horkheimer, M. 1969. *Dialectic of Enlightenment*. London: Verso.

Bankoff, G. 2001. Rendering the World Unsafe: 'Vulnerability' as Western Discourse. *Disasters*, 25(1): 19–35.

Bell, D. 1973. *The Coming Post-Industrial Society*. London: Heinemann.

Berridge, V. 2007. Multidisciplinary Public Health: What Sort of Victory? *Public Health*, 121(6): 404–408.

Botha, D., Van Niekerk, D., Wentink, G., Coetzee, C., Forbes, K., Maartens, Y., Annandale, E., Tshona, T. and Raju, E. 2011. Disaster Risk Management Status Assessment at Municipalities in South Africa. Research Report: African Centre for Disaster Studies. North-West University, Potchefstroom. Online: http://acds. co.za/uploads/research_reports/Salga_Draft_ReportFINAL_V1_2_small.pdf (Accessed 3 May 2013).

Buzan, B., Weaver, O. and de Wilde, J. 1998. *Security: A New Framework for Analysis*. Boulder: Lynne Rienner Publishers.

Doornbos, M. 2003. Good Governance: The Metamorphosis of Policy Metaphor. *Journal of International Affairs*, 57(1): 3–17.

Douglas, M. 1991. *Purity and Danger: An Analysis of the Concepts of Pollution and Taboo*. London: Routledge.

Escobar, A. 1988. Power and Visibility: Development and the Invention and Management of the Third World. *Cultural Anthropology*, 3(4): 428–443.

Foucault, M. 1982. The Subject and Power. *Critical Inquiry*, 8(4): 777–795.

Friedman, T. 2006. *The World Is Flat: The Globalized World in the Twenty-First Century*. London: Penguin.

Fukuyama, F. 1989. The End of History? *The National Interest*, 16: 3–18.

Harvey, D. 2005. *The New Imperialism*. New York: Oxford University Press.

Laurie, N., Adolina, R. and Radcliffe, S. 2005. Ethno-development: Social Movements, Creating Experts and Professionalizing Indigenous Knowledge in Ecuador. *Antipode: A Radical Journal of Geography*, 37(3): 470–496.

Merton, R. K. 1938. Social Structure and Anomie. *American Sociological Review*, 3(5): 672–682.

Rogerson, C. M. 2010. In Search of Public Sector-Private Sector Partnerships for Local Economic Development in South Africa. *Urban Forum*, 21(4): 441–456.

Smith, A. 2005. *An Inquiry into the Nature and Causes of the Wealth of Nations*. Hazleton: Pennsylvania State University Press.

South Africa. 2002. *Disaster Management Act. Act 57 of 2002*. Pretoria: Government Printer.

South Africa. 2005. *National Disaster Management Framework*. Pretoria: Government Printer.

Wisner, B., Kent, G., Carmalt, J., Cook, B., Gaillard, J. C., Lavell, A., Oxley, M., Gibson, T., Kelman, I., Van Niekerk, D., Lassa, J., Willison, Z. D., Bhatt, M., Cardona, O. D., Benouar, D. and Narvaez, L. 2011. Political Will for Disaster Reduction: What Incentives Build It, and Why Is It So Hard to Achieve? A Contribution to the Review of the Draft Global Assessment Report 2011, Chapters 5, 6 & 7. Draft 7b. Unpublished paper.

1 A genealogy of disaster risk assessment and management

Introduction

A more comprehensive account of DRAaM is required before any critical analysis of the South African industry can commence. This chapter offers an historical account of DRAaM's development and institutionalisation globally by using Foucault's genealogical method. As such, the analysis is not meant to reveal linear progression or regression. Genealogies are not based on (unrealistic) commitments to clear indisputable causal linkages. Rather, genealogy traces the emergence of discourse by explaining which discourses emerged with it, perhaps even preceding it at times, though sometimes also occurring in parallel. Associated discourses set the necessary conditions for a particular discourse to develop in the way it has, by creating a particular frame of reference which facilitates the discursive and other practices leading to, in this case, DRAaM's particular institutionalisation. The chapter traces DRAaM's development along two broad discourses; those which emerged in development and those pertaining to another apparent contemporary global metanarrative or macro-discourse, being risk assessment and management.

As the emergence of discourse is essentially mediated by power, genealogy seeks to reveal that what is often viewed in a positive light by those exercising power may also be to the detriment of others. The following analysis reveals that disaster reduction (as the field was initially known), as an idea with initial emancipatory objectives, has been appropriated by the bureaucratising influence of instrumental reason. It has essentially been co-opted by forces focusing on the top-down implementation, formal institutions and structures, as opposed to the substantive positive change initially envisaged. By framing danger associated with disaster in the idiom of risk assessment and management, DRAaM has emerged as an essentially conservative discourse.

The development discourse

During the 1970s the idea that a disaster is a *force majeure* and purely the function of *hazards* (or looming triggers) exogenous to social life was increasingly challenged. In particular, at the forefront of this initiative was a research unit based at the University of Bradford in the UK consisting of a group of young geographers. These authors famously asserted that we must 'take the naturalness out of natural disasters' (see Westgate et al., 1976). According to them there are no natural disasters, merely natural hazards. Disasters result from human interaction with hazards. Authors writing from this general viewpoint essentially argued that disasters were a function of socio-economic *vulnerability*. In other words, some people are more susceptible to the adverse effects of natural or manmade hazards on account of living in floodplains or in highly flammable informal housing arrangements. Those more susceptible are most often the poor. Even where the wealthier segments of the population are affected, their privilege implies that they can afford adequate insurance. The implication of this position was clear and made explicit: there is a clear link between disasters and development.

In theory, disasters are both a consequence of development and a potential cause of underdevelopment. The basic logic is that patterns of development often associated with poor planning render populations vulnerable to the adverse effects of hazards. Disasters may also affect development by destroying infrastructure. They may disrupt livelihoods for example by destroying crops or rendering inaccessible the roads used to travel to work. Cavallo and Noy (2010:30) provisionally find that poorer countries are more likely to be adversely affected; therefore effective developmental interventions are required to limit the adverse effects of disasters on development. To gain further insight into the role of development discourse in the institutionalisation of DRAaM, the role of international policy conventions and discourses of sustainability should be scrutinised more closely.

Sustainable development

The Brundtland Report titled *Our Common Future* (WCED, 1987) is an appropriate starting point for discussion on the relationship between international conventions and DRAaM. The legacy of this report is observable in numerous contemporary discourses in some way linked to or developed under the guise of sustainable development. The sustainable development discourse has also provided the context for, among others, corporate social responsibility, climate change mitigation and environmental rights to develop. The 1987 report defines sustainable development as development today which does not jeopardise future development. According to this definition the implications for disaster reduction are quite clear, though DRR was by no means a global institution at that point. *Our Common Future* laid the groundwork for subsequent high-profile conventions, such as the Earth Summit in Rio in 1992 and the adoption of Agenda 21 or the Rio Declaration and the establishment of the Commission on Sustainable Development and the World Summit on Sustainable Development (WSSD) in Johannesburg in 2002, which culminated in the well-known Millennium Development Goals (MDGs, see UN, 2002). A new set of sustainable development goals (SDGs) were adopted to succeed the MDGs 2015. The Rio Declaration stressed such issues as linking development and the environment, conserving natural resources, limiting pollution, meeting basic (human) needs, addressing population growth and managing urbanisation. The MDGs and the subsequent SDGs have a similar focus, but with a greater focus on basic human needs. Conventions stemming from the climate change conferences in Copenhagen, Durban and Paris may be viewed similarly, or rather as part of the continued influence of sustainable development discourse. As McEntire (2000:34) notes, the World Commission on the Environment and Development (WCED) drew a lot of attention to a reflexive vision of development and hence, modernity's adverse consequences. The report defined sustainable development and simplified its central tenets, hence its major impact. McEntire argues that the presence of the sustainability discourse facilitated the development of the DRAaM discourse, or DRR, as he more conventionally articulates it.

Up to the late 1980s and the Brundtland report, DRR was mostly an academic debate. By 1989 however, disaster reduction, as it was still known, was formalised in the form of the International Decade for Natural Disaster Reduction (IDNDR). Other DRR conventions followed through the 1990s and the 2000s, including the Yokohama Strategy and Plan for a Safer World (IDNDR, 1994), the Hyogo Framework for Action 2005–2015 (HFA) (UNISDR, 2005) and the Sendai Framework for Disaster Risk Reduction 2015–2030 (SFDRR) (UNISDR, 2015). The United Nations resolution A/RES/44/236 (UN, 1989), signed in 1989, established the IDNDR. This initiative was both a set of policy objectives and an organisation established within the UN Secretariat and headquartered in Geneva, which was tasked with overseeing the implementation of the aforementioned policy objectives. By the end of the 1990s it was decided that there should remain a UN organ dedicated to DRR, though its name changed to the United Nations International Strategy for Disaster Reduction (UNISDR). Since the late 1980s the field unfolded in part through the stipulations of conventions. There have been many consistencies indicated by shared words and phrases. These include emphases on science, technical information, risk assessment and information dissemination awareness and education for risk reduction (cf. IDNDR, 1994:5, 7, 14; UNISDR, 2005:8; UNISDR, 2015:14). Much of this language focussed on technical skills that likely aided a context conducive to professional 'expert' risk assessors to be contracted for this purpose. The Yokohama Strategy for a Safer World (hereafter the 'Yokohama strategy'), for example, states that risk assessment is a required step towards adoption of 'adequate and successful disaster reduction policies and measures'. Risk assessment is explicitly cited a number of times, though also by implication through scenarios, forecasting and management. Furthermore, risk assessment along with early warning formed one of five priorities for action (UNISDR, 2005:2).

A second matter repeatedly emphasised is capacity development. The word 'capacity' appears 24 times in the 25-page HFA. Unfortunately, for the most part this capacity building merely manifests as formal state structures and developing policies at national or lower levels of government. Yokohama Strategy states the need for 'national plans' and 'national committees'. This theme is carried through to the Sendai Framework, which, excluding where it is part of a name, includes more than 35 references to plan, plans or planning in its 37 pages.

Notions of risk assessment and management, though arguably always dominated by interpretations favouring practitioner or elite-level activity, have seemingly become so over the years. Local participation and local or traditional knowledge is variously acknowledged. However, De la Poterie and Baudion (2015) observe an alarming shift in emphasis from community actors as important partners to recipients of risk information. Linked to the technocratic portrayal of the problem through technical assessment and formal planning is a dominant cultural conception of danger and disaster. In this regard, the distinction McEntire (2004:195) draws between two broad approaches to disaster research is informative. The first body of research is largely based and informed by the aforementioned cultural approach and cultural explanations of risk. As such it assumes that society should be educated or made aware in order to better understand the risk they face and therefore to avoid risky behaviour. The second body of literature follows a more structural and critical approach, emphasising the material forces that limit a household's, group's or geographic area's capability to protect itself (e.g. Middleton and O'Keefe, 1997). Through clear associations with cultural interpretations of danger, the official message seems to be largely that the contextually dispersed adversities of disaster remain to be addressed through education and other forms of social engineering, replacing ignorance with a 'culture of safety'.

Concepts such as 'culture of safety' serve to illustrate the point that these conventions present a fertile environment for generic and instrumentalised rationalities to predominate, based upon performance indicators distilled from essentially vacuous truths. Accordingly, failure to alleviate the afflictions of disaster can always be attributed to such empty yet forceful signifiers by some or other degree of extension and malleable interpretation, also encrypted in the official document. The emphasis on 'political commitment' (cf. IDNDR, 1994:10; UNISDR, 2015:9) is indicative of the emerging governing discourses of the time, whereby accountability seems to be reserved only for national or lower levels of government in the Global South, under the guise of good governance. Global governance and international regimes which potentially undermine, directly or indirectly, the safety of ordinary citizens are not treated in the same manner. The truths propagated as a result of power-appropriating DRR instructs us that often profound threats can simply be managed away, at least to a significant extent. Consequently, in the contemporary context of neoliberal governmentality informed by notions of stripped-down, efficiency-oriented neoliberal management the globally emergent DRR that focuses on policies and capacity building (read new state institutions) may easily become fetishised performance indicators, definitive of DRAaM.

Perhaps to their credit, some delegates attending the World Conference on Disaster Reduction in Kobe, Japan, in 2005 recognised how more general developmental goals and objectives for DRR are linked. This is why some have argued that the Millennium Development Goals (MDGs) and the HFA are mutually reinforcing (e.g. World Bank, 2011:32). A similar sentiment is reiterated in the Sendai Framework. Based on this logic, DRR cannot occur to any meaningful extent if the MDGs are not attained. Some commentators have however labelled the MDGs unrealistic (Clemens and Moss, 2005). However, what those linking the MDGs with the HFA have achieved yet again is to clearly link poverty and danger, though once again not taking the argument to its full completion by largely and conveniently ignoring or at least downplaying the fundamentally political underpinnings of this relationship. If poverty, defined in accordance with the more precise MDGs – though still problematic for its fetishism of 'one-size-fits-all' performance indicators – were to link with DRR to the point that the MDGs became a prerequisite for DRR, then this notion of poverty could not merely be skimmed over. Analyses of danger have to be critical of social marginalisation (and its mirror image) and the singular modernity which has and continues to produce both.

Global sustainability discourse has also influenced academic discussion. As is explained in the introduction to the current chapter, categorically proving causality is an impossible undertaking. Some conjectural inferences may nonetheless be drawn. The often-used Sustainable Liveli-hoods Framework (SLF), as a tool for micro-level analyses, did emerge within four years of the Brundtland report. Practitioners and academics often use the SLF in analyses of disaster. As subsequent discussions indi-cate, it has also been reworked into conceptual frameworks specifically developed to study disasters.

Chambers and Conway (1992) produced a seminal text in which they proposed *Sustainable Livelihoods: New Concepts for the 21st Century*. The SLF has since become an often-used tool in the development community for analyses, especially for rural contexts. It appears that SLF thinking not only chronologically follows from sustainable development thinking but is rooted in exactly the same logic of protecting assets in order to mitigate shocks. The SLF and the Access Model bare a significant resemblance, sug-gesting that the former significantly influenced the latter.

The Pressure and Release and Household Access Models

The two most-often-used theoretical models to explain disaster risk are those devised by Blaikie et al. (1994) in their book *At Risk: Natural Hazards, People's Vulnerability, and Disasters*. The Pressure and Release Model and the Household Access Model venture to explain the so-called progression of vulnerability and how hazards and vulnerability interact at the house-hold level. Vulnerability in the disaster studies literature is often treated as closely related to poverty, although authors often add the qualification that these two concepts are not identical (Wisner et al., 2003:11). Whereas it is the typically marginalised members of society who are the most vulnerable, poverty tells us very little about the causal chain between vulnerability and risk. It might therefore be asked, exactly how is it that members of society often categorised as impoverished tend to be most at risk?

The Pressure Model traces the progression of vulnerability from root causes, such as an authoritarian regime, global trade relations and so on to dynamic pressures, such as migration, which eventually lead to unsafe condi-tions whereby individuals or households are vulnerable to particular hazards. Unsafe conditions might for example entail living in a particularly hazard-prone area. Figure 1.1 verifies the linear logic this model rests upon.

The objective of DRAaM accordingly is to reverse the progression of vul-nerability. Therefore, this figure may theoretically be turned around so that the arrows point in the opposite direction, from unsafe conditions towards root causes. This would be the release component within the Pressure and Release (PAR) Model, which implies combining both graphs, progression left to right and release, right to left. A Foucauldian interpretation of reality

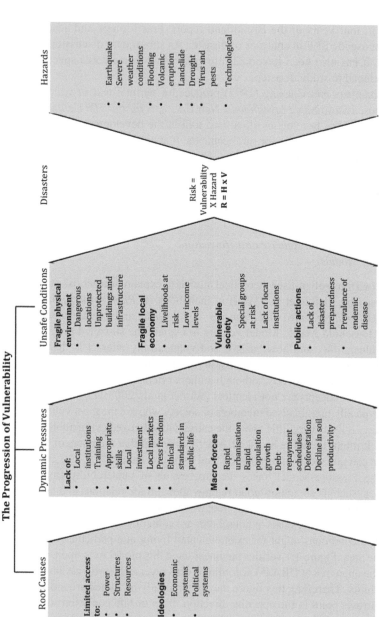

Figure 1.1 The Pressure and Release Model: the progression of vulnerability

would dismiss this model immediately upon an initial cursory glance. The PAR Model, however, deals with complex social reality with overly simplistic reductionism, ignorant of the central defining characteristic of society that is structuring power.

To address the matter of inappropriate linearity, the authors added a second model, more explicitly focused on the household level. The Household Access model presents disasters, hazard and vulnerability as a cycle. It ventures to explain how each cycle of hazard and vulnerability interaction produces a new iteration, which again feeds into processes of vulnerability. The Access Model emphasises the role of household assets (economic, social, physical, natural) in reducing or increasing vulnerability throughout

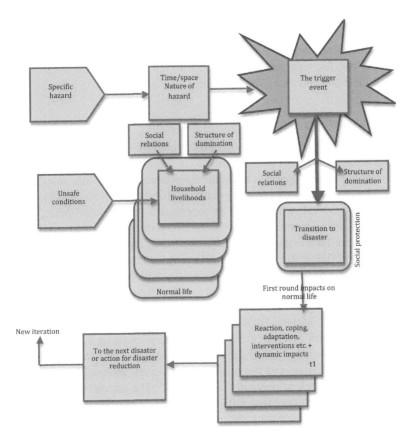

Figure 1.2 The Household Access Model

Source: From Wisner et al. (2003)

the iterative process of hazard and social, political, economic and physical environmental interaction. As such, it is similar to the SLF if explicitly applied to disasters.

The two analytical models presented by Wisner et al. (2003) might be construed as a compromise between accuracy and usefulness. The reason for claiming that these authors' work constitutes a compromise is the fact that they acknowledge structural issues or root causes that Marxist scholars would typically emphasise while still equally focusing on micro-level issues which might be subject to interventions by those involved in the development set. Naturally, this has led to criticism from both sides on the political spectrum, a debate I do not have sufficient space to comprehensively deal with here. However, a general point of this book is that discourses of DRAaM, including those on vulnerability, serve to preserve the status quo. They are conservative, by way of legitimising an increasingly emergent professional agenda on the part of those privy to the DRAaM code. These individuals and organisations use tools such as PAR and Access Models which are, for their generic vagueness in the absence of supplementary analytical content, exceptionally blunt tools.

Institutionalising a 'lack of progress with DRR'

The approach prescribed through policy conventions on DRR is that of mainstreaming, a term which seems to continually crop up in development circles.[1] Where reference is made to 'mainstreaming DRR into sustainable development', it appears to mean that responsibility is devolved to the local government level, where DRM is coordinated across departments, each contributing its particular expertise to risk sensitive development. There have been efforts to measure the effectiveness of DRM in these terms, both by the UNISDR and civil society (cf. GNDR, 2009; UNISDR, 2011). These reports seem to indicate progress, especially with priority national awareness and two risk assessments, but at the same time having especially poor results regarding the priority of addressing underlying risk factors. This seems logical. 'Making DRR a national priority' and getting practitioners to conduct a risk assessment would surely be a lot easier than addressing the 'root causes' of prevalent dangers. The global master plan, that of mainstreaming, is failing, and this mostly is blamed on 'incapacity', 'bad governance' or 'a lack of political will for DRR'. The aforementioned global reviews of performance with DRR also include cases of developed-country local government incapacity. Incapacity is not purely a developing-country problem, though it may be more acute in contexts where there are fewer resources. Mainstreaming, however, most likely compounds the problem. If local government capacity is a general problem in most parts of the world, then what do we expect when we add the additional and overwhelming weight of mainstreaming DRR into sustainable development at the local level? Even where both these highly pregnant and vague concepts are given explicit context-specific meaning, mainstreaming, by implication, requires new institutions and coordination across sectors. Such a cumbersome risk management approach, dictated through generic policy, implies altering the allocation of resources, even where disaster risk might not be the most urgent need. It is therefore perhaps not surprising that Lewis and Kelman (2012) argue that while initiatives such as the global assessment reports can be commended, top-down DRR has not worked.

We must now consider the second major set of discourses facilitating DRAaM's emergence: the global metanarrative of risk assessment and management. While adding to the preceding discussions, the following section will conclude by solidifying the conclusion of the current section presented in the previous paragraph.

The idiom of risk assessment and management

Risk assessment and management as a mode of analysis and decision making have been part of particularly Western discourse for a very long time. The genealogy of risk assessment and management can be traced back to the development of the probability theory. A few prominent intellectuals during the 17th and 18th centuries, such as Laplace and Pascal, are credited with making some of the original strides in the field. Gambling had a lot to do with probability analysis's emergence, in particular the game called craps. Pascal, for example, is said to have solved 'the dice problem' puzzling gamblers of the time. This problem can be articulated as follows: How often must one throw two dice to have a greater-than-even chance of throwing two sixes? Cardona solved the earlier dice problem for a single die in 1525 (Ore, 1960:413). Pascal is also credited with having solved the other major puzzle of his time, the so-called division problem. This problem is noted in Italian manuscripts as early as 1380 and, according to Ore (1960:414), is likely of Arabic origin. The problem in brief entails the division of prize money in a tournament interrupted before its natural completion. In such cases the question that remains is what is the probability to win for each participant?

Some argue that the advent of modernity, from the 17th century, greatly facilitated the spread of this knowledge (Sheynin, 1977:204), although the practice of insuring property can be traced to 2000 BC, when participants in trade caravans in the Near East agreed to share damages in case of robbery. Such risk pooling was also found in maritime commerce in Ancient Greece (Sheynin, 1977:206). The same author traces (disability) allowances paid to guild members in cases of incurable disease back to 1284 and annuities back to the 14th century (Sheynin, 1977:210). At that stage the practices were not yet subjected to the sophisticated calculations they are today. What has indeed happened since the arrival of modernity is that more and more methodologies for risk assessment were developed in diverse fields. These fields include the economic and management sciences, environmental impact assessment, epidemiology, engineering, biology, criminology, occupational health and safety, agriculture and the insurance industry. It can thus be argued that given the prevalence of risk assessment and management, this idiom provided a convenient framing also for disaster reduction towards the end of the 1980s. The post-1980s era of commodified expert knowledge, broadened and intensified by the end of the Cold War, presented the opportunity for the expert risk assessor and management consultant to emerge and be schooled and employed in disaster risk management. Sociologists such as Ulrich Beck (1992) and Anthony Giddens (1990, 1999) have written on the increasing fixation with risk and, by extension, risk assessment and management in Western society. These sociologists formulated the so-called risk society thesis, for which Beck is especially famed.

The risk society and risk-based decision making

Beck offers a theory of Western modernity in which he suggests that a fixation with common threats and their management is the primary characteristic of that society. He argues that the welfare state's success has significantly alleviated class cleavages. At the same time, levels of education and technological advancement have increased risks and perceptions of risk. Western modernity is therefore obsessed with risk, and the relationship between individuals and hazardous technologies has become the most significant form of struggle in what he deems 'World Risk Society'.

Giddens (1990, 1999) offers similar commentary, though framing the matter slightly differently. He argues that risk is a consequence of modernity. According to him, risk emerged in the Middle Ages. Giddens argues that there does not seem to be any notion of risk in what he deems 'traditional' cultures and that the defining characteristic of the risk society is a 'forward looking mentality'. Giddens warns that we should not dwell exclusively on the negative consequences of modernity pertaining to risk. The negative and positive sides of risk, such as the benefits drawn from hazardous technologies, are not separate from one another (Giddens, 1999:10). The risk society is preoccupied with 'reflexive modernisation', which contrasts with 'simple' modernisation. Reflexive modernisation according to Giddens (1999) is how we deal with the 'limits and contradictions of the modern order'.

One might argue that the risk society thesis in its current guise(s) is not particularly relevant to developing world contexts. This, however, does not mean that it and other similar discourses have not directly or indirectly impacted developing world contexts and governing discourses. The dominant Western academic perspectives on diverse dangers seem to prescribe risk management as a type of reflexive modernisation and thereby refuse to challenge the structural causes of danger, which are part of the very DNA of this modernisation. The idiom of risk assessment and management with roots in mathematics by penetrating various fields over the years appears to have been a readily available and convenient narrative to attach to disaster reduction once the limits of disaster response became apparent.

Some authors within the social sciences have previously offered critical analyses of risk assessment and management. Wildavsky (1981:19), commenting within the USA context, for example, asks whether assessments are 'worth the amount we are paying for them'. There might be significant opportunity costs involved, meaning that the funds might have been spent elsewhere to more meaningful effect. Furthermore, he argues that very little significant action follows from these assessments in any event. Wildavsky therefore validates a previous argument that a 'lack of capacity' or deficient 'political will' and a lack of implementation or any similar notions are not limited to a particular part of the world, nor are they typically confined to the Global South. Others have argued similarly. Rayner (2007:166) notes, 'The process of assessment has become a substitute for meaningful but inevitably controversial intervention'. In other words, the paper trail left by institutionalised assessment offers a safe means of spending a budget and consequently has often become the key performance indicator in the applicable fields. It seems that Western or Western-influenced societies are using 'science' to make political decisions without adequately considering the political implications for society. This scientistic technocracy comes at the expense of participatory decision making, while the entire risk assessment and management process may be essentially undemocratic.

Much of contemporary interventions are rooted in notions of culture and perception and work by psychologists focusing on individual risk perception, which is contrasted with 'objective' risk. This literature often portrays humans as ignorant and voluntarily engaging in risky behaviour. The logical requisite intervention is therefore education. Oltedal et al. (2004:6), however, argue that there is very little evidence supporting this type of cultural theory of risk. Moreover, based on arguments presented already, positions linking safety to education already contain within them the prospect of being ethnocentric. The conceptual materials offered by many cultural theories of risk are therefore exceedingly dangerous tools. Douglas (1991) warns against such approaches, suggesting that rational behaviour is always risk averse. Settling in a floodplain, for example, may be the only option a household has in order to achieve geographical proximity to job opportunities or other resources.

Sociologists responded critically to this analytical approach, contrasting perception with the 'reality' followed by many psychologists (Tierney, 1994). Their critique was more or less that the social construction of risk does not affect 'at-risk' populations, exclusively. Simply put: risk assessors' claims to objectivity are fallacious. More and 'better' information will not solve these value disputes. Though contested values seem a rather euphemistic notion, Tierney's reading is similar to Rayner's previously cited argument that dangers and the contrasting perspectives on danger are essentially political issues. These contested values are manifestations of political struggle, past and present. For this reason, to contrast risk perception with objective risk serves very little purpose. These are simply different perspectives, a product of the multiplicity of truths and realities found in different geographical and temporal spaces. By largely relying on expert-driven approaches to ascertain risk, the de-politicising and re-politicising functions of discourse are recast as the values of some segments of society which are by implication considered less legitimate and subjected to the values of others who have access to overriding contemporary knowledges. In the case of DRAaM, up to this point, one obvious omission in the institutionalisation of the risk assessment and management idiom remains. The requisite mathematisation was duly added in the early 1990s, albeit 'pseudo' mathematisation.

During the early 1990s Blaikie et al. (1994) published their seminal book *At Risk: Natural Hazards, People's Vulnerability, and Disasters*, cited earlier. The book introduced the then-new conceptual framework for analysing disaster risk as presented in the PAR and Access models discussed previously. The book also introduced what the authors call a 'pseudo equation'. Once again it should be emphasised that causal linkages are difficult to prove in historical analyses. Nonetheless, this equation presented a neat representation of DRAaM theory in what might be viewed as an idiom of risk assessment and management based on techno-mathematical attempts to control nature. The equation is presented as follows (Wisner et al., 2003:49):

$$R = H \times V$$

Others have slightly altered this equation. Nonetheless, it does appear as though the following version is most common in discussions on DRAaM (Lautze et al., 2005:9):

$$R = H \times V \div C$$

In this scheme Risk (R) is equal to Hazard (H) times Vulnerability (V) divided by Capacity (C). The concepts of hazard and vulnerability have already been explained. Capacity, also referred to as coping capacity, denotes the various ways in which people are able to withstand the adverse effects of hazards. Coping capacity refers to how households are able to draw on these resources in order to protect themselves against a hazard. Just as vulnerability implies an increase in susceptibility, capacity implies a decrease and the ability to withstand or lessen the impacts of hazards.

Even though Blaikie et al. (1994) add the disclaimer, calling it a pseudo-equation, the economic, technical, positivist risk assessment and management idiom is evidently encoded in this representation of the disaster problem. Similar to the assets and capital discussed earlier, this formula essentially defines social relations in economic/managerial vernacular. These conceptual tools, therefore, by definition, entail reification by taking social relations and reframing them in a way that largely obscures these underlying social and political relations. This equation, one might argue, forms the basis for or at least potentially sets the scene for risk assessment as a technical exercise executed from a distance. This distance may be physical or simply the ontological gulf emerging from a lack of meaningful dialogue with those in the process reduced to the objectified assessed.

Conclusion

This chapter provided a brief genealogy of DRAaM traced along two broad strands of literature. These are development discourse and the discourse of risk assessment and management. Even though disaster reduction might initially have been a noble idea holding the promise of emancipation, it has been corrupted by instrumental reason. This fundamentally political matter essentially linked to social relations is framed in terms of top- down bureaucratised implementation through generic performance indicators based on international policy convention and technical risk assessment and management jargon. Leveraging such jargon within the frame of risk assessment and management is a de-politicising act, as it does not acknowledge the need for political solutions to political issues.

It is also re-politicising by reassembling the status quo on the terms of the already enfranchised expert practitioner and, as we have learned, to little avail. Disaster reduction subjected to salient discursive powers of the past 40 years has, as a result, unfolded as an essentially conservative discourse, inappropriately regimented and aligned with a ubiquitous, economistic technique from the management sciences. In the following chapter the focus shifts to South Africa specifically. It explains the disaster problem by drawing on DRAaM vernacular before exploring an alternative conception of disaster in Chapter 3.

Note

1 Mainstreaming is a concept that has been used in development in many different contexts. It seems to roughly refer to considering or giving precedence to one issue while working on another. Therefore one might mainstream HIV/AIDS into food security interventions or climate change adaptation into DRR or DRR into sustainable development or mainstream gender into development. All of the potential overlapping instances of mainstreaming cause much confusion.

References

Beck, U. 1992. *Risk Society: Towards a New Modernity.* London: Sage.

Blaikie, P., Cannon, T., Davis, I. and Wisner, B. 1994. *At Risk: Natural Hazards, People's Vulnerability, and Disasters.* London: Routledge.

Cavallo, E. and Noy, I. 2010. The Economics of Natural Disasters: A Survey. Inter-American Development Bank Working Paper Serious, Number IDB-WP-124. Online: http://www.iadb.org/res/publications/pubfiles/pubIDB-WP-124.pdf Date of access: 23 January 2012.

Chambers, R. and Conway, G. 1992. Sustainable Rural Livelihoods: Practical Concepts for the 21st Century. IDS Discussion Paper 296. Institute of Development Studies, University of Sussex, Brighton.

Clemens, M. and Moss, T. 2005. What's Wrong with the Millennium Development Goals? Centre for Global Development Brief. Online: www.cgdev.org/files/3940_file_WWMGD_pdf (Accessed 6 May 2016).

De la Poterie, A. T. and Baudion, M. 2015. From Yokohama to Sendai: Approaches to Participation in International Disaster Risk Reduction Frameworks. *International Journal of Disaster Risk Science*, 6: 128–139.

Douglas, M. 1991. *Purity and Danger: An Analysis of the Concepts of Pollution and Taboo.* London: Routledge.

Giddens, A. 1990. *The Consequences of Modernity.* London: Polity Press.

Giddens, A. 1999. Risk and Responsibility. *The Modern Law Review*, 62(1): 1–10.

Global Network of Civil Society Organisations for Disaster Reduction (GNDR). 2009. Views from the Frontline. Online: http://gndr.org/fr/programmes/views-from-the-frontline/vfl-2009.html (Accessed 5 May 2016).

International Decade for Natural Disaster Reduction (IDNDR). 1994. Yokohama Strategy and Plan of Action for a Safer World: Guidelines for Natural Disaster Prevention, Preparedness and Mitigation. Geneva: IDNDR.

Lautze, S., Aklilu, Y. and Boyd, E. 2005. Assessments & Appeals: Strengthening Non-food Emergency Responses in Ethiopia. The Livelihoods Program: Saving Lives and Livelihoods. USAID. Online: http://www.who.int/hac/crises/eth/sitreps/Lautze_et_al_Assessments_and_Appeals_2005Final.pdf. Date of access: 14 February 2014.

Lewis, J. and Kelman, I. 2012. The Good, The Bad and The Ugly: Disaster Risk Reduction (DRR) versus Disaster Risk Creation (DRC): Public Library of Science (PLoS) Currents: Disasters. 21 June. Online: http://currents.plos.org//disasters/article/the-good-the-bad-and-the-ugly-disaster-risk-reduction-drr-versus-disaster-risk-creation-drc/pdf (Accessed 23 September 2013).

McEntire, D. A. 2000. *From Sustainable to Invulnerable Development: Justifications for a Modified Disaster Reduction Concept and Policy Guide*. Doctoral dissertation. University of Denver.

McEntire, D. A. 2004. Development, Disasters and Vulnerability: A Discussion of Divergent Theories and the Need for Their Integration. *Disaster Prevention and Management*, 13(3): 193–198.

Middleton, N. and O'Keefe, P. 1997. *Disaster and Development: The Politics of Humanitarian Aid*. London: Pluto.

Oltedal, S., Moen, B. E., Klempe, H. and Rudmo, T. 2004. Explaining Risk Perception: An Evaluation of Cultural Theory. Trondheim: Norwegian University of Science and Technology. Online: ftp://131.252.97.79/Transfer/ES_Pubs/ESVal/risk_perception/Cultural_theory.pdf (Accessed 5 February 2014).

Ore, O. 1960. Pascal and the Invention of Probability Theory. *The American Mathematical Monthly*, 67(5): 409–419.

Rayner, S. 2007. The Rise of Risk and the Decline of Politics. *Environmental Hazard*, 7(2): 165–172.

Sheynin, O. B. 1977. Early History of the Theory of Probability. *Archive for History of Exact Sciences*, 17: 201–259.

Tierney, K. J. 1994. Sociology's Unique Contributions to the Study of Risk. Paper presented at the 13th World Congress of Sociology. 18–23 July, Bielefeld, Germany. Online: http://udspace.udel.edu/handle/19716/589 (Accessed 5 February 2014).

United Nations. 1989. General Assembly Resolution A/RES/44/236. International Decade for Natural Disaster Reduction. 22 December. Geneva: United Nations.

United Nations. 2002. General Assembly Resolution 56/195. International Strategy for Disaster Reduction. 21 January. Geneva: United Nations.

United Nations International Strategy for Disaster Reduction (UNISDR). 2005. Hyogo Framework for Action 2005–2015. Geneva: UNISDR.

United Nations International Strategy for Disaster Reduction (UNISDR). 2011. Hyogo Framework for Action 2005–2015 Mid-Term Review. Online: http://www.unisdr.org/we/inform/publications/18197 (Accessed 5 May 2016).

United Nations International Strategy for Disaster Reduction (UNISDR). 2015. Sendai Framework for Disaster Risk Reduction 2015–2030. Geneva: UNISDR.

Westgate, K., O'Keefe, P. and Wisner, B. 1976. Taking the Naturalness out of Natural Disasters. *Nature*, 260(5552): 566–567.

Wildavsky, A. 1981. Richer Is Safe. *Financial Analysts Journal*, 37(2): 19–22.

Wisner, B., Blaikie, P., Cannon, T. and Davis, I. 2003. *At Risk: Natural Hazards, People's Vulnerability and Disasters*. London: Routledge.

Wisner, B., Kent, G., Carmalt, J., Cook, B., Gaillard, J. C., Lavell, A., Oxley, M., Gibson, T., Kelman, I., Van Niekerk, D., Lassa, J., Willison, Z. D., Bhatt, M., Cardona, O. D., Benouar, D. and Narvaez, L. 2011. Political Will for Disaster Reduction: What Incentives Build It, and Why Is It So Hard to Achieve? A Contribution to the Review of the Draft Global Assessment Report 2011, Chapters 5, 6 & 7. Draft 7b. Unpublished paper.

World Bank. 2011. *The Sendai Report: Mainstreaming Disaster Risk Management for Sustainable Development Managing Disaster Risks for a Resilient Future*. Washington DC: World Bank.

World Commission on Environment and Development. 1987. Our Common Future. Published as Annex to UN General Assembly Document A/42/427. Online: http://www.regjeringen.no/upload/SMK/Vedlegg/Taler%20og%20artikler%20av%20 tidligere%20statsministre/Gro%20Harlem%20Brundtland/1987/Address_at_ Eighth_WCED_Meeting.pdf (Accessed 23 January 2012).

2 South Africa's disaster risk profile

Introduction

The current chapter and the next provide a foundational description and analysis, which render a particular conceptualisation of the status quo regarding prevalent dangers. As such they provide a benchmark for comparing dominant DRAaM practices during the course of Chapters 4, 5 and 6. While Chapter 3 explicitly spells out this conceptualisation, the current chapter is delivered at a more basic level, from a point of departure at the heart of the common epistemological ground that this book shares with the DRAaM industry. However, this is merely a point of departure, and analyses in this chapter and the next gradually depart from the modes of description and analysis more typical of DRAaM. This chapter is focused more on the materiality of danger, while the following chapter adds to this foundational work through analyses of associated, less tangible, yet profoundly significant forms of suffering.

Navigating South Africa's disaster risk profile poses certain methodological challenges. Reliable disaster statistics are notoriously difficult to find in most parts of the world. Though many databases exist, they are all limited in terms of scope and/or quality of information. These databases draw on many different definitions of the concept of disaster and therefore often provide very different data. This hampers both accuracy and comparability, whilst the fact that not all disaster events are reported further adds to the data problem. South Africa is no exception. There is no comprehensive country-level database. Hence, the following analysis cannot discuss the exact prevalence of specific hazards and their geographic distribution across the country, as it is essentially confined to previous studies of disasters and danger in more particular locales. Certain areas have been researched more extensively than others. This may be because these areas are more disaster prone. However, proximity to a DRAaM–focused research institution likely plays an equally significant role, as is the case with the Western Cape Province.

The Research Alliance for Disaster and Risk Reduction (RADAR, formerly known as Disaster Mitigation for Sustainable Livelihoods Programme or DiMP) based at Stellenbosch University has, since its inception in the late 1990s, developed a substantial regional body of DRAaM knowledge, especially on urban fires and floods.

The literature reviewed in the current and following chapters covers research sites in the Western Cape, the Northern Cape, the North-West Province, the Eastern Cape and Gauteng. These DRAaM–framed studies are supplemented by analyses from related bodies of literature, such as housing, migration, HIV/AIDS and many more, which collectively speak to the national level. Salient features of the country's risk profile have been distilled qualitatively, through thematic data analysis.

The argument presented in the current chapter is that the material dangers associated with vulnerability and risk are to a large extent a result of political-economic forces, rooted in South Africa's predominantly racialised capitalist modernity. It is a modern experience in which race and social class overlap decidedly. The analysis challenges cultural explanations of risk as a tool for meaningful change by emphasising the structural drivers of danger.

Overview

South Africa is not a particularly disaster-prone country. However, many smaller though more frequent events do occur (Holloway, 2009). Most South African disaster literature focuses on urban as opposed to rural dangers. While this practice may generally be agreed with, it can also be criticised. South Africa has a roughly equal rural and urban population. It is indeed true that danger is mostly an urban phenomenon. As such, this largely urban focus seems appropriate, especially when one considers the upsurges in urbanisation since apartheid formally ended. The subsequent concentration of largely unemployed residents or informal settlements often causes susceptibility to hazards such as flooding, strong winds and urban fires. Limited available land may cause extremely dense settlements, often partially constructed of highly flammable materials such as wood and where flammable materials are a common source of energy. The result is often that fires, once started, spread very quickly. The same dearth of inhabitable land may also force some residents to settle on floodplains or even in river beds. Having made these general comments on the most prevalent features of urban danger, it should be noted that migration does not imply that threats to life and livelihood are negligible in rural areas.

The current analysis of prevalent dangers explores prominent themes from both urban and rural hazards and the associated dynamics of exposure (vulnerability), as discussed in the DRAaM literature. It must however be emphasised that such categorisation is necessarily an exercise in oversimplification. Not only does space not allow for in-depth analysis of each and every local context, but the literature remains thin in terms of in-depth context-specific analysis. This limitation is striking, as such an emphasis on vulnerability as purportedly *the* defining characteristic sets DRR/DRM apart from previous response-orientated approaches to disaster management. The rest of this chapter includes discussions on a limited selection of hazards and matters pertinent to these hazards. Many of the dynamics rendering someone susceptible to a particular threat are also relevant to various other threats discussed in the chapter, though such duplication is avoided in order to make the most of the limited available space.

Informality

Disasters are attracted to liminal spaces. Those caught up in the transitory positions within the social structures of society tend to be relegated to transitory or non-permanent physical spaces. So-called informal settlements are the quintessential South African liminal spaces. This is where liminality takes on permanence. In post-apartheid South Africa transformation has meant perennial limbo rather than renewed access to the political and economic spoils of society. It is indeed incomplete liberation. We shall return to this politics of transition in the next chapter. For immediate purposes however, in light of this link between liminal spaces and hazardous or dangerous spaces, it should be no surprise that some hazards have even been described as afflictions of informality.

Fires in informal settlements are commonly mentioned in disaster risk assessments, research reports and the academic literature. According to Pharoah (2009:112), fire risk is primarily caused by socio-economic conditions. Urbanisation, as previously noted, often brought on by rural unemployment, places great stress on urban government and urban service delivery. The logical result of such immense urbanisation is that the need for resources often far exceeds availability. At the same time, urban employment opportunities are not created at a sufficient rate. According to StatsSA (2016), 27.7% of all South Africans remain unemployed in terms of the strict definition of the concept, which includes only those actively seeking employment. This figure rises to 36.3% when the expanded definition of unemployment is applied, including all citizens of working age not employed and not or no longer actively searching for work. By way of logical deduction, if South Africans migrate to find employment but unemployment continues to rise, then clearly migration, at least on aggregate, does not alleviate the prevailing lack of income-earning opportunities. In addition, many of those who are employed remain poor. With little or no income and financial resources, new entrants to urban areas are left with no other option than to occupy generally hazardous living conditions. This often means that residents of informal settlements and therefore the typical victims of informal settlement fires make use of what Schwebel et al. (2009) refer to as alternative fuel sources for cooking and/or heating, such as paraffin, wood or coal. The sheer prevalence of these energy sources, however, hardly suggests them being an alternative. Compounding the various causalities of hazard and exposure is another common characteristic of informal settlements. Some homes are built with flammable materials such as wood, and very often they are erected very close together. Consequently, once a fire starts, it can spread easily and quickly. Therefore, time frames are often too narrow for meaningful response. I know from personal conversations with officials and firemen and women over a number of years that many homes, even an entire settlement, can burn down before the fire brigade even arrives. This is not on account of tardy response. It often takes less than 10 minutes for homes to be razed to the ground.

To address the problem of settlement fires Schwebel et al. (2009) recommend more awareness campaigns. These authors do add a qualification, acknowledging the fact that more knowledge does not always lead to safer behaviour, and thus they also recommend legislative regulation of paraffin-fuelled appliances. But awareness and legislation might not be feasible interventions to reduce urban fires. Settling in an area where there is very little space available, making use of flammable building materials and using alternative energy sources such as paraffin heaters may all be rational decisions, even if we assume there is always a choice to be made between one or more viable options. These suggested interventions and the assumptions they are based on are reminiscent of the previously mentioned cultural explanations of disaster. In contrast to such cultural theories we might argue that there are opportunity costs to choices involving safety. These may include living very far away from places of (potential) employment or not being able to afford your children's school uniforms. The likely compromise to accommodate these expenses might include the type of paraffin heater acquired and (flammable) building materials used in home construction. Most significant, however, are the bondages enforced by a generally marginal existence, which many ordinary South Africans face daily. These bonds inform the multifarious ties binding the diverse power-laden causalities pertaining to urban fires. This violence, found in the everyday mechanics of life, is the focus of the following chapter. For now though, the following quotation by MacGregor et al.

(2005:23) should illuminate this important and complex dynamic, to be revisited in due course:

> [I]t must be noted that the triggers in themselves may not be the primary cause of fire since social problems such as domestic violence and alcohol abuse can be considered as predisposing factors for events such as a candle toppling.

If practitioners were to use the cultural approach to disaster and 'teach' people not to drink, often done in the context of extreme destitution, they might be considered rather naïve, if not hypocritical, and committing the same labelling discussed in the introductory chapter. Substance abuse has largely been medicalised in mainstream popular and scientific discourse and interventions. Why then does this not hold equal sway for the Other, who, should we continue this cultural theory–informed thought experiment, will be yet again portrayed as inferior on account of inadequate education? While the suburbs are full of psychologists with no or little African language proficiency, these less affluent South Africans' problems would, to take assumptions of cultural theories to their logical conclusion, once more be equated to their individual and implied reprehensible agency.

Urban flooding

Many parts of South Africa experience frequent and severe flooding, especially parts of the Western Cape Province. Areas affected by severe storms and flooding include the west coast, the southern Cape and the City of Cape Town (CoCT) (Tempelhoff et al., 2009). Based on available literature some general observations are presented.

Urban floods have become a frequent problem in the southern Cape, especially in the Eden District Municipality (EDM). This municipality includes urban areas, such as George, Knysna and Oudtshoorn. The southern Cape coast's near-annual exposure to extreme weather compounds urbanisation pressures, the consequences of which are reflected in costly losses to the less financially secure households that often inhabit runoff and flood-exposed areas. Urban agricultural production also suffers regular flood losses on account of damaged transport and other essential infrastructure.

As is the case with urban fires, the causes of urban flooding are numerous and diverse. Rapid urban growth increases the prospects of flood losses by adding to the space covered in unabsorbent hard areas, compacted through increased pedestrian traffic over a decreasing number of possible walkways (Tempelhoff et al., 2009). This decrease in water absorption also increases runoff and places water management infrastructure under pressure (DiMP,

2007). Bare and compacted areas essentially speed up narrower but more voluminous water streams, which simultaneously erode the uncovered soil. The situation is exacerbated by the destruction of wetlands due to urban development, as former wetland areas no longer absorb surplus water. Settling in floodplains obstructs the flow of water, exacerbating the extent to which water backs up on river banks. Mgquba and Vogel (2004), in reference to Alexandra Township near Johannesburg, emphasise flood losses due to the significant concentration of people in the area. While the population grows, newer entrants are forced to settle in lower-lying areas.

Increased urban growth since apartheid has not been met by significant improvements in water management infrastructure. Maintenance and continued development of such infrastructure have not kept up with rapid urbanisation. Many studies cite blocked water drainage infrastructure as a resultant cause of flooding (cf. Benjamin, 2008:59; DiMP, 2007; Tempelhoff et al., 2009). These blockages force run-off water back into the streets. The losses inflicted by floods can be harsh to individual households, settlements and key industries, especially in a national context in which unemployment is a significant crisis.

The southern Cape is a major tourism region, and that industry tends to be affected severely and often (Tempelhoff et al., 2009). The consequences are enormous in monetary terms, though this observation requires significant qualification. For example, in 2006 the southern Cape area was affected by two storms, leading to significant financial loss. But there are also non-tangible losses. Examples in this regard include important documents, such as identity documents (IDs), being lost or ruined, children missing school due to wet uniforms, and fatalities. Economic statistics also mask the fact that members of low-income households sometimes miss work and hence part of their wage income due to blocked access routes (DiMP, 2005:68). In such cases the consequences are much more specific and concentrated in terms of lived-through experiences than general statistics on economic losses would suggest. Some households and individuals might even be forced into chronic (income) poverty. In this regard a critical point must be reiterated. Residents of particularly hazard-prone areas are regularly affected by a number of cumulative perils. For example, when sewerage systems are flooded, this flooding may trigger or coincide with diseases, such as tuberculosis, influenza, respiratory tract infection, body rashes or cholera. Furthermore, flood-affected citizens typically lack sufficient funds to spend on private medical care, while accessible state-funded medical facilities are under-resourced (Ataguba and Akazili, 2010).

As with urban fires, urban flooding most commonly affects residents of low-lying 'informal' homes. Additionally, floods tend to cause severe damage to poorly constructed 'RDP homes' (named after the Reconstruction and Development Plan under which this housing project was introduced on the early 1990s), even to the extent that these houses are uninhabitable. Post-apartheid RDP housing projects tend to be outsourced. Even though all formal (brick-and-mortar) homes have to comply with a set of standards, many contractors are not held to these standards. Given such leeway, these contractors use cheaper and inferior building materials and thus construct substandard structures. Poor planning in some instances compounds the problem of poor building quality when houses are built in low-lying areas with an already high water table (Benjamin, 2008:145). There are exceptions to the relationship between being flood affected and social class, though these are mostly negated through access to insurance.

Following the 2006 southern Cape floods, the losses for those living in informal settlements and RDP houses were indeed much worse. Benjamin (2008:148) estimates losses based on household interviews. He reports that flood losses meant that it took informal settlement residents between one and three years to recover financially. Those residing in RDP houses took at least four months to recover from the flood. Benjamin (2008:148) also notes that household members would often have innovative ideas on how best to weatherproof a dwelling. They were however constrained by a lack of financial resources and time (Benjamin, 2008:162). Therefore, once again in contrast to cultural conceptions of disaster, it is not that these households lack knowledge or are flawed in some or other way. The more likely explanation is that residents lack the financial/material means to implement their ingenuity. Settling in hazardous conditions or remaining there after flood losses are not necessarily decisions people make easily.

As with other urban dangers, people often settle in flood-prone areas, as the land is free, close to an access route to (potential) employment or they 'have nowhere else to go'. Even in cases where the threat is clear and others' homes have been destroyed, Mgquba and Vogel (2004:34) found that residents still decided not to leave. Those already affected contemplated acquiring materials to build new homes, but, once again, there were opportunity costs. When making such decisions Mgquba and Vogel (2004:35) note that in the aftermath of the 2001 floods in Alexandra Township, the choice for some residents was between rebuilding their homes and paying for their children's education. Thus, the evidence seems once again to contradict cultural notions of risk, as these 'choices' appear to be based on rather reasonable assessments of the present while duly considering the future. One final example should suffice in illustrating the limited options many South Africans have in avoiding the danger of flood-afflicted areas.

The area colloquially known as the Cape Flats on the outskirts of Cape Town is annually affected by severe flooding.[1] In this instance, physical geography plays a role, as these 'flats' offer no natural slope conducive to proper drainage. However, the human geography of the matter is arguably more significant. Informal settlements, such as those on the Cape Flats, are often found in such low-lying or otherwise marginal areas. In the case of Cape Town there is no other uninhabited land in relative proximity to the mere promise of employment held by the central business district and industrial areas. Those with the means to do so live in the suburbs and drive to work. Others simply cannot afford that option.

Communicable disease

The disaster studies literature tends to classify epidemics as hazards or slow-onset disasters. That being said, there is little literature dealing with the matter as such in the South African context. One example is the work by Burger and Brynard (2001), who argue that HIV/AIDS is a creeping or slow-onset disaster or a silent epidemic. It is possible for a person or group to be vulnerable to HIV and AIDS due to inadequate nutrition, gender and associated inequalities and the previously regulated migrant labour systems, which forcibly separated husbands and wives by way of influx control policies. This splintering of households in a context of gender inequality logically increased the number of the husbands' sexual partners, who on account of gender are believed to be generally more entitled. Other underlying causes include unemployment and poverty. HIV/AIDS feeds back into marginalising processes due to the particular working age group who typically fall prey to this condition (Burger and Brynard, 2001:178). Furthermore, HIV/AIDS impacts society much like other disasters do. It has devastating effects on the national economy, with massive social consequences throughout society, such as children growing up without parents. It also impacts social welfare and medical aid schemes (Burger and Brynard, 2001:176), potentially exceeding vast proportions of society's ability to cope.

In addition to the effects of the HIV/AIDS pandemic, though closely related to it, is South Africa's very high incidence of tuberculosis. South Africa accounts for 5% of global incidences (Brundage et al., 2011). According to Janse van Rensburg-Bonthuyzen (2005:192) treatment is not sufficiently effective due to, among many factors, the concomitant HIV/AIDS pandemic and poorly resourced state medical facilities. Consequently the prevalence rate increases every year, as tuberculosis has a high comorbidity with HIV/AIDS. As TB is the leading cause of death for people with HIV, treating it effectively will probably go some way towards alleviating the effects of the HIV/AIDS pandemic. Such a simple correlation does not yet reveal the stark realities of epidemiology. If one takes an approach to poverty based on income, which is still the norm, the links between poor public health, little or no income and living in informality are obvious, even though there might be many qualifications to these correlations. Informal settlements tend to be densely populated. These liminal spaces often lack readily available water and sanitation facilities, which significantly exacerbate the spread of disease. In addition, burning coal in the winter as a source of heat creates smoke, which, combined with cold and wet conditions, increases the risk of upper-respiratory-tract infections. Finally, diseases like cholera and gastro-intestinal infections result from contaminated drinking water, which is of inconsistent quality in South Africa.

Mining accidents, sinkholes and acid mine water

South Africa is not particularly prone to earthquakes. Only one noteworthy such event has occurred during the past 50 years. This took place in the Western Cape in 1969. Nevertheless, mining activity in Gauteng, the North-West and Free State provinces continually causes numerous smaller tremors and sinkholes (Van Eeden et al., 2003:101). These sinkholes result from the relationship between mining and the dolomitic areas where it often takes place. Much of Gauteng and most of the gold mining areas in the West Gauteng are underlain by dolomite. Groundwater often seeps through these porous geological structures into gold mines. For many years mines simply pumped the water out to access gold reefs more easily. As water levels in some of the natural ground water compartments declined, sinkholes formed. Sinkholes first appeared in the 1950s. With the controlled influx of migrant workers, the urbanisation of residents, the new planned housing settlements and growing informal settlements, these hazards became more significant.

Van Eeden et al. (2003:120) argue that the prevalence of sinkholes in an area also has psycho-social effects on communities. In other words, the threat of sinkholes and their impact on physical well-being may lead to emotional distress. With the exception of some parts of the City of Tshwane, in particular the suburb of Centurion, the primary beneficiaries, either directly or by some degree of extension, of South Africa's mineral wealth need not be threatened by sinkholes. Instead, this danger is found in the historically proximate labour reserves, often strategically placed on the opposite side of the buffer offered by open land or industry. Here marginalised South Africans remain in harm's way above the ground, just as they remain below the ground daily and often fatefully in the process of earning a wage. Based on insights from previous analyses, it could be argued that later entrants to these marginal settlements most likely chose the area as the land was free. In addition, this increasingly marginal land would have been close to work, therefore saving the unaffordable expenses of commuting. Besides sinkholes and frequent fatalities, mining as a longstanding engine of South Africa's modernity has another legacy, often framed as disastrous. The legacy in question is acid mine water and industrialisation's effects on the country's fresh water reserves. The same mining techniques, which altered the water table, exposed groundwater to a variety of minerals. This continues to cause heavy metal contamination and acidification of water destined for some of the country's drinking water reserves (Adler et al., 2007:34). However, conventional water purification methods cannot remove many of the heavy metals, including the uranium in some cases, found in water reserves.

Some blame the laissez-faire approach to mining by the state as a major cause (Adler et al., 2007:33). Nonetheless, Funke et al. (2007) explain that the situation is complicated by the fact that mining houses are unwilling to accept responsibility. Acid mine drainage and related problems such as sinkholes are largely the consequence of their predecessors' collaboration with the apartheid government in order to sustain mining enterprises during an era of extensive sanctions against the apartheid state. The state, as regulatory authority, could hardly afford to sanction these enterprises. Bills had to be paid and foreign currency generated, and gold was the answer. Other concerns, such as safe and sustained acid water treatment and confinement for long-term public health, were inconveniences not immediately pertinent.

The Vaal River is perhaps the most salient example of a polluted South African water course. This river, as a major deposit for industrial waste from south Gauteng and the north of the Free State Province, poses potential threats to communities located downstream and activities in and around the river related to agriculture and tourism. While water pollution and the resultant public response potentially threaten state legitimacy, industry continues to pollute water and modify the water table. Adler et al. (2007:33) cite further potential losses, for example in foreign direct investment, social unrest and racial tension, as access to formal water delivery follows the familiar contours of social class. As such, even this fundamental prerequisite for life very much remains racialised. While perception and material realities may at times diverge, some are able to allay their fears by digging into their pockets to purchase bottled water. This is merely an addition to the modern politics of the binary underlying categories such as formal and informal living. The latter sometimes implies no on-site water whatsoever.

Risk assessment reports mention additional urban hazards and disasters. These additional hazards and disasters include transportation accidents, air pollution and technological disasters, such as industrial accidents. The latter is a case in point of the fact that most matters related to DRAaM remain vastly contested. Industrial accidents might not directly affect informal settlements more than other areas. However, writing within environmental sociology, Cock (2004) takes a different position, arguing that it is often the working-class or unemployed citizens who are located closest to industrial areas and therefore most likely to be affected by the adverse consequences of the ensuing pollution. Furthermore, industrial accidents might have a more detrimental impact on working-class incomes, as these households' main breadwinners, in contrast to managers for example, will more likely be in direct contact with hazardous work environments.

Rural hazards

As agriculture is the most significant livelihood activity in rural South Africa, rural areas often most acutely experience the effects of drought. Rural flooding may be experienced as similarly severe by farmers. Floods cause crop failures, in particular in the unique winter rainfall area of the Western Cape. For the purposes of the current study, however, it would be more meaningful to explore the droughts that are experienced more broadly and for which there is more available literature.

South Africa is a relatively dry country, and the situation seems to be worsening. As the country's demand for water exceeds its supply, fresh water sources deteriorate. Even though droughts are part of the normal South African agricultural reality, they might increase due to contemporary and future manifestations of climate change (CDKN, 2014). Drought means far more than merely the absence of rainfall. It also relates to the level of total moisture in reservoirs, rivers and the soil. Drought should be viewed in context. Different types of agricultural activities require different amounts of water. Similarly, different contexts have over many years developed their own forms of resilience, more or less susceptible to a relative anomaly. The same absence of rainfall will, for example, have far more profound effects on crop production compared to livestock rearing.

As previously noted, about half of South Africa's population is rural and dependent on agriculture. Many people residing in former homeland areas are subsistence farmers on marginal land. This is largely the result of the 1913 Land Act, which essentially enforced environmental degradation by segregating the majority of the population to 13% of the land. Land was worn out over time by cattle-raising practices, more sustainable under previous land distributive regimes. Many farmers still depend on these continuously degrading natural resources for their livelihood. To this day, mostly white (commercial) farmers are much better resourced than (black) communal farmers and are better equipped to absorb the shocks inflicted by drought. White farmers have generally had the obvious advantage of capital accumulation and the resilience one associates with relative wealth on larger pieces of land, while they could regularly rest parts of their land to ensure sustainability. This resting is generally facilitated by greater amounts of boreholes spread out over the farm, again a result of larger pieces of land and the potential for capital accumulation it provides.

Many rural areas also suffer damage from virtually annual veld (field or grassland) fires. Veld fires are essentially a problem during the dry season. For most of South Africa this is during winter. The only exception is the winter rainfall region of the Western Cape Province. In this province, occasional extreme heat exacerbates the dry summers and thus the potential for veld fires to erupt. Although fires usually occur during dry seasons, they still require a trigger. Potential triggers are arson, lightning, honey hunters, fires started in order to create fire breaks which then escalate, power lines falling down and smokers who indiscriminately dispose of cigarette butts.

The implications of veld fires are severe for farmers losing crops, animals and other assets and, more generally, the environment. Wildfires can lead to considerable hydrological and geomorphological change, even indirectly. With the loss of plant life caused by fires, the bedrock is exposed and begins to weather through exposure to winds and possible water runoff. As a result, the structure of the soil changes due to wind erosion and landslides or other debris flows (Shakesby and Doer, 2006). Thus, in addition to the immediate acute effects these fires have on agriculture, there are also cumulative effects over time, degrading agricultural land.

Veld fires also occur in urban areas. As with most other disaster-related dangers, those living on the fringes of the city, who are often migrants and who often live in informal settlements, are typically most affected. To reiterate, living on the liminal, still-to-be-fully-integrated margins of a metropole is dangerous. The fact that peri-urban and urban areas steadily expand into open grassland places residents in harm's way. These locales of informality bear many of the previously mentioned exacerbating characteristics, such as alternative fuel sources and temporary homes fashioned from flammable materials, built close together. Consequently this expansion simultaneously increases the chance that veld fires erupt.

Conclusion

This chapter provided a disaster risk profile for South Africa. In the process I have argued that 'vulnerability' and 'risk' are to a large extent a result of political-economic forces, rooted in South Africa's predominantly racialised capitalist modernity, where race and social class overlap decidedly. Given this reality, cultural explanations of risk, typically operationalised through public education and awareness campaigns, offer a tool for meaningful intervention that is limited at best. The material dangers associated with disaster likely require more fundamental interventions into the structural process characterising everyday life. However, material realities are intertwined with less tangible social realities and forms of suffering. The following chapter builds on this one by conceptualising this entanglement of the material and the less tangible. This conceptualisation of the everyday offers a reference point from which the institutionalisation of DRAaM in South Africa is critiqued in Chapters 4, 5 and 6.

Note

1 The Cape Flats refers to the low-lying, relatively flat (and sandy) area on the outskirts of Cape Town, where many low-cost formal and informal settlements are found. The area is perennially subject to winter floods.

References

Adler, R. A., Claassen, M., Godfrey, L. and Turton, A. R. 2007. Water, Mining, and Waste: An Historical and Economic Perspective on Conflict Management in South Africa. *The Economics of Peace and Security Journal*, 2(2): 33–41.

Ataguba, E. J. and Akazili, A. 2010. Health Care Financing in South Africa: Moving towards Universal Coverage. *CME*, 28(2): 74–78.

Benjamin, M. A. 2008. *Analysis of Urban Flood Risk in Low-cost Settlements of George, Western Cape, South Africa: Investigating Physical and Social Dimensions*. Master's thesis. University of Cape Town.

Brundage, S. C., Bellamy, W. M., Bliss, K. E., Canfield, S., Cooke, J. G., Lamptey, P., Morrison, J. S. and Reeves, M. 2011. Terra Nova: How to Achieve Successful PEPFAR Transition in South Africa. A Report of the Global Health Policy Center. Washington DC: Center for Strategic and International Studies (CSIS). Online: https://csis.org/publication/terra-nova (Accessed 14 February 2014).

Burger, D. and Brynard, P. A. 2001. Disaster Management Perspectives and Challenges into the New Millennium. *Journal of Public Administration*, 36(2):169–181.

Climate and Development Knowledge Network (CDKN). 2014. The IPCC's Fifth Assessment Report: What's in It for Africa? Online: cdkn.org/wp-content/uploads/2014/04/AR5_IPCC_Whats_in_it_for_Africa.pdf (Accessed 24 May 2016).

Cock, J. 2004. Connecting the Red, Brown and Green: The Environmental Justice Movement in South Africa. Research Report. Centre for Civil Society/Centre for Development Studies, University of Kwazulu-Natal: Pietermaritzburg. Online: http://ccs.ukzn.ac.za/files/Cock%20Connecting%20the%20red,%20brown%20and%20green%20The%20environmental%20justice%20movement%20in%20South%20Africa.pdf (Accessed 23 January 2012).

Disaster Mitigation for Sustainable Livelihoods Programme (DiMP). 2005. Disaster Briefing December 2004 Cut-off Low: Overberg, Cape Winelands, Central Karoo and Eden District Municipalities. Cape Town: Disaster Mitigation for Sustainable Livelihoods Program (DiMP). University of Cape Town.

Disaster Mitigation for Sustainable Livelihoods Programme (DiMP). 2007. Severe Weather Compounds Disaster: August 2006 Cut-off Lows and their Consequences in the Southern Cape, South Africa. Cape Town: Disaster mitigation for Sustainable Livelihoods Programme (DiMP). University of Cape Town.

Funke, N., Nortje, K., Findlater, K., Burns, M., Turton, A., Weaver, A. and Hatting, H. 2007. Redressing Inequality: South Africa's New Water Policy. *Environment: Science and Policy for Sustainable Development*, 49(3): 10–23.

Holloway, A. 2009. *Disasters, Risks and Development: Integrated Perspectives.* Honours Course Outline. University of Cape Town.

Janse van Rensburg-Bonthuyzen, E. 2005. Staff Capacity and Resources at Nine Free State Clinics: Shortcomings in the TB Programme. *Acta Academica Supplementum*, 1: 192–220.

MacGregor, H., Bucher, N., Durham, C., Falcao, M., Morrissey, J., Silverman, I., Smith, H. and Taylor, A. 2005. *Fire Hazard and Vulnerability Assessment for Informal Settlements: An Imizano Yethu Case Study with Special Reference to the Experience of Children.* Cape Town: Disaster Mitigation for Sustainable Livelihoods Programme (DiMP): University of Cape Town.

Mgquba, S. K. and Vogel, C. 2004. Living with Environmental Risks and Change in Alexandra Township. *South African Geographical Journal*, 86(1): 30–38.

Pharoah, R. 2009. Fire Risk in Informal Settlements in Cape Town, South Africa. In Pelling, M. and Wisner, B. (eds.) *Disaster Risk Reduction: Cases from Urban Africa.* London: Earthscan. pp. 103–125.

Schwebel, D. C., Swart, D., Hui, S. A., Simpson, J. and Hobe, P. 2009. Paraffin-related Injury in Low-income South African Communities: Knowledge, Practice and Perceived Risk. *Bulletin of the World Health Organ*, 87(9): 700–706.

Shakesby, R. A. and Doer, S. H. 2006. Wildfire as a Hydrological and Geomorphological Agent. *Earth-Science Reviews*, 74(3–4): 269–307.

Statistics South Africa (StatsSA). 2016. Unemployment Increased in the First Quarter of 2016 (May). StatsSA, Media Release-Quarterly Labour Force Survey.

Tempelhoff, J. W. N., Van Niekerk, D., Van Eden, E., Gouws, I., Botha, K. and Wurige, R. 2009. The December 2004–January 2005 Floods in the Garden Route Region of the Southern Cape, South Africa. *Jàmbá: Journal of Disaster Risk Studies*, 2(2): 93–112.

Van Eeden, E. S., De Villiers, B., Strydom, H. and Stoch, L. 2003. Effects of Dewatering and Sinkholes on People and Environment: An Analysis of the Carletonville Area in Gauteng, South Africa. *Historia*, 48(1): 95–125.

3 Writing structural violence

Conceptualising ordinary South African being

Introduction

The material realities of disaster and danger presented in the previous chapter are intertwined with also less tangible, yet very powerful, forms of suffering which many South Africans experience. Acknowledging these intersectionalities significantly aids a case for more substantive social change, which extends far beyond popular conceptions of risk reduction. This chapter argues that to mitigate disaster also requires actions that do not take disaster as the primary foci. At the same time, disasters are often not residents' most pressing concern. Danger is fundamentally political and urgent, and it is part of a problem far more vast, chronic *and* acute, than a framing such as DRAaM allows for.

For many daily life might best be described as occurring under conditions of structural violence, where society's very (dis)functioning reproduces material and less tangible forms of suffering. The implication is that meaningful, large-scale social change, the type promised by the 1994 democratic transition but unfortunately not produced for many ordinary citizens, should be foregrounded in any analysis of the contemporary South African political economy. I do not suggest that there was no large-scale social change post-1994; far from it. However, there have been numerous continuities and many new forms of suffering, while new iterations of previous adversities have been – and continue to be – reproduced through the dialectics of South Africa's democratic transition. The critique presented here consistently keeps this dialectic in the background, from time to time nudging it to the fore to highlight issues central to the argument.

Structural violence is not exercised directly by one person on another but by a structure produced and reproduced through custom, law or politics in general (cf. Degenaar, 1980; Galtung, 1969). In this regard Allan Boesak referred to 'the structural violence of apartheid' institutionalised through discriminatory laws and enforced by the police force of the day (Degenaar,

1980:19–20). This chapter argues that structural violence is still a prominent feature of contemporary South African society, even if it manifests somewhat differently and is sustained through different structures. Structural violence has material properties, such as danger, infections and hunger and less-material components such as the psychological distress imposed by material hardship and denigrating public discourse. In order to present the narrative of structural violence, the discussion is arranged around a set of themes. The analysis is historically informed but not chronological. These headings serve the basic purpose of structuring the argument. They are neither exhaustive nor do they represent discrete categories. The analysis also highlights and emphasises patterns and intersectionalities of suffering.

The dialectics of transition

Apartheid was officially institutionalised from 1948. Intellectually it hinged upon a particular brand of applied anthropology, *Volkekunde* (which roughly translated is the study of people, as in 'a people' or *volk*). Based on the so-called ethnos theory, it represented ethnic groups as discrete entities with certain innate biological and cultural traits. Each supposedly had its own nature and therefore needed its own space to flourish in 'its own way'. *Volkekunde*, along with the findings of various commissions of inquiry, were meant to justify the state's practices of social control. These endeavours included addressing such issues as (Afrikaans) white poverty in the form of the Carnegie Commission of 1932 and later the various commissions aimed at regulating the presence of black South Africans in designated white areas. As Thomas (2010:117) notes, 'The commission reports have a recurring theme of Africans as rooted in rural, collectivist lifestyles'. It was the 'duty of paternalistic whites to preserve these identities, or each distinct group's "ethnos" '. In practice, this translated into an extensive state bureaucracy first known as *Naturelle Administrasie* (Native Administration) and later as *Ontwikkelings Administrasie* (Development Administration). Development Administration was even institutionalised as a separate academic discipline. Official doctrine was maintained through clandestine academia and a massive state bureaucracy, subjecting populations to the vagaries of management through what Gordon (1988:550) refers to as a 'science of social control'. Apartheid therefore entailed a type of instrumental reason aimed at maintaining racial and Afrikaans superiority and might as such be viewed as a previous iteration in a process of dialectical change.

As explained, globally the period roughly since the late 1970s has coincided with the emergence of neoliberal governmentality. It is since then that globalisation accelerated through liberal trade policies, Structural Adjustment Policies and the growth of the post-industrial economy.

The post-apartheid state clearly bought into many aspects of this type of thinking, through its self-imposed structural adjustment programme, often associated with the package of economic policies contained in the Growth Employment and Redistribution (GEAR) policy framework. The policies hatched under the GEAR framework were often self-contradictory. Inflation targeting, for example, came at the expense of economic growth. In addition, fiscal restraint came at the expense of redistribution, as the latter was sought through trickle-down growth, facilitated by foreign investment. It is now general knowledge that this investment never materialised to any significant extent. Bond (2005) offers an extensive analysis of South Africa's post-apartheid economic policy. He outlines how adopting neoliberal economic policies is a result of elite transition. In essence, the governing elite became increasingly convinced of the merits of neoliberal economic policy. Post-apartheid elites were co-opted by way of neoliberal discourses that were encouraged by institutions such as the World Bank and the International Monetary Fund, even though South Africa was never formally subjected to structural adjustment. Nonetheless, the aforementioned neoliberal policies have and continue to administer a modernity which is coded with conservatism in its very DNA. The result has been at worst an increase in poverty and inequality or at best a lack of improvement in this regard. The official poverty rate is 56.8% based on a 2008–2009 survey, while income inequality, in the form of the Gini coefficient, is 0.7 (StatsSA, 2016).[1] This means that South Africa is consistently ranked amongst the most unequal societies, if not the most unequal society, in the world. The post-apartheid status quo is characterised by the coexistence of a small, previously advantaged (still mostly white) and recently empowered black elite and an impoverished (overwhelmingly black) majority. The transition from authoritarianism to democracy might therefore best be described as one oligarchy replacing another, with significant overlap in the beneficiaries of both dispensations. South Africans have felt and continue to experience various forms of suffering on either side of the 1994 transition. The remainder of this chapter deals with these dimensions and intersectionalities of suffering.

Segregation

Segregation has been a key vehicle for the oppression of the majority of South Africans throughout the 20th century. Though typically associated with the Afrikaans word 'apartheid' (separateness), segregation did not emerge with apartheid. Space only allows for meaningful discussion of the period since the beginning of the 20th century, even though this phenomenon may be traced back much further. In attempts to consolidate white dominance, the self-governing Union under the British Crown passed the 1913 Land Act. Accordingly a mere 13% of land was allocated to the black majority. In later years the country would produce numerous pieces of legislation aimed at separating whites from blacks. These measures ranged from the Group Areas Act of 1950 to the Bantu Homelands Citizenship Act of 1970 to the Pass Laws Act of 1952 to the Immorality Act of 1927 and Prohibition of Mixed Marriages Act of 1949. The latter two essentially implied that it was immoral for individuals from different races to have sexual intercourse. Similarly, the Pass Laws Act determined that all black South Africans were required to carry a passbook when moving in white areas. The book, among other personal details, contained the name of the individual's white employer and the duration of his/her employment, essentially serving as a valid reason for being in the white area. The Bantu Homelands Citizenship Act determined that black Africans were no longer South African citizens but rather citizens of quasi-independent homelands. These policies were typically rationalised based on the quasi-scientific principles discussed earlier and the notion of 'good neighbourliness' or 'separate but equal'. This of course is a contradiction. By its very nature, with proclivities of non-association towards a particular category of people, the implication is that that population group is not deemed worthy of association and therefore not equal. What essentially took place was that white and black South Africans were socialised into believing that they were fundamentally different. This, for example, manifested as petty apartheid, whereby different races had different entrances to shops and separate benches in parks, to name but two examples. The country was riddled with plaques in public areas which read 'Whites only'.

The associated humiliation, distress and the accompanying material consequences for black South Africans can be found in the South African literature of the past half century. Here I draw on literature to reveal the delicate and intricate meanings and emotions and as such consequences of structural violence. One such example is Adam Small's work *Kanna hy kô hystoe* (1999 edition), a play dealing with the misfortunes of being *coloured* (of mixed race) in apartheid South Africa. The main character, *Kanna*, returns home after many years abroad to attend his adopted mother *Makiet's* funeral. His adopted family had years earlier recognised his potential and tried to optimise his chances for professional success and upward social mobility by spending much of their extremely limited resources on his education. This success he did find, overseas, where he became an engineer. The drama plays out as a set of recollections of instances of torment. One such instance is the scene where *Diekie*, Kanna's younger brother, is defenceless in front of a magistrate as he is sentenced to death by hanging for taking revenge upon his sister *Kietie*'s husband *Poena*, who sold her into prostitution. The line: *'as hulle vir Diekie versagting gevind het, moet hulle vir duisende versagting vind'* (if they found mitigating circumstances for Diekie, they would have to find mitigating circumstances for thousands) is subsequently repeated. The suggestion here seems to be that assigning blame to those bearing the brunt of structural violence is somewhat problematic. Socialisation into a violent and depraved day-to-day existence often produces individuals who produce violence themselves. This is evidenced, for example, by the proportion of members of minority groupings incarcerated (Gordon et al., 2012).

Many other literary works written on the unjust nature of apartheid appeared under that regime. One such example is the biography of Poppie Nongena (*Die Swerfjare van Poppie Nongena*), written by Elsa Joubert and first published in 1978. The work tells the story of a woman of mixed race married to a black man. She is sent from pillar to post by the apartheid bureaucracy under the practices of racial classification, forced removals and the homeland policy. This trapped her in a perennial state of liminality as a result of insecure tenure and lack of available living wage–paying jobs. Neither she, nor, towards the end of the book, her children could escape the structural violence of the system and the physical violence in the country which ensued.

The migrant labour system implemented before and during apartheid was part and parcel of the segregating practices of the time. Single-sex hostels were built in urban areas, where black men, mostly working on the mines, could live. Women were not allowed to accompany their husbands to urban areas. Rather, they stayed in the homelands, typically engaging in small-scale agricultural activities, on the 13% of the land designated to black South Africans. Since the 1880s the central economic thrust of South Africa's capitalist modernity was upheld by the country's mineral wealth. This was accompanied by worker exploitation through inadequate wages. As Wolpe (1972) notes, paying miners minimal, below-subsistence-level wages was precisely feasible because of the existing pre-capitalist economy in homeland areas. Workers did not need to be paid sufficient wages for family subsistence, as these livelihoods were already subsidised by rural subsistence farming. Wages could be kept low, facilitating the process of (local and global) white capital accumulation. Where black women did urbanize, they were often employed as domestic workers, raising white children, often at the expense of the emotional neglect of their own. The pittance they were paid either freed up productive time for white women to add a second household income or it facilitated a life of leisure (Cock, 1980).

Education

As with many other aspects of society, apartheid has left an enduring impression on the South African education system. Schools and tertiary education institutions, as with all other facets of public and private life, were segregated under apartheid. To a large extent this is still the *de facto* case, though the system-legitimising discourse has changed. *Bantu Education* was the name given to the black education system during apartheid. Black South Africans were taught a different curriculum, in accordance with 'their particular talents'. This often precluded them from professional occupations. During the mid-1970s the government changed certain policies so that education in important subjects, such as mathematics, would be taught in Afrikaans only. As a result, the inexorable tensions plaguing the country were ignited, spectacularly culminating in the 1976 Soweto uprising and the killing of schoolchildren by the police. Segregation in education was not limited to the country's school system only.

Universities were also segregated. One institution, the University of the Western Cape (UWC), was designated for coloureds. The University of Durban-Westville (UDW) was reserved for Indians, and each homeland (read each ethnic group) had its own 'bush college'. White South African tertiary institutions were merely divided between (the majority) Afrikaans and English universities. White universities were far better resourced than those designated for other groups. Change has been limited since the advent of democracy. Some universities merged as per ministerial directive, and most Afrikaans universities, at least to a predominant extent, became English-medium institutions. Former white universities now have many more black students, though the opposite is most certainly not true. Tuition fees have, however, increased significantly since apartheid, as the proportion of university income from state subsidies has declined significantly. This has caused nationwide protests amid escalating student debt. There are formidable challenges to access in a society where many households cannot even afford to service the interest on student debt. There are also concerns with a student culture on many campuses shaped by whites for whites and which many black students experience as hostile or exclusionary.

The primary and secondary school system has also seen limited transformation. Even though segregation is no longer official policy, once again the *de facto* case seems to be that the formerly white schools, which remain far better resourced and staffed than schools in townships, are the purview of the middle class, who more often than not are white. The situation is exacerbated by many qualified South African teachers choosing to immigrate to the United Kingdom in particular, where the majority of foreign teachers are South Africans. This exodus occurs amid the fact that the higher education system annually produces too few newly qualified teachers (SACE, 2011:6–7). Of course upper-middle-class or exceptionally (typically athletically) gifted children have exclusive access to private or semi-private/semi-state schools. These schools often have exorbitant fee structures or at least charge fees that are out of the reach of the overwhelming majority of households.

Employment

South Africa's skills base does not compare well internationally. The education system seems to be failing in this regard. Grade 12 pass rates are low. Unfortunately, those who do obtain university exemption are often ill prepared. There is a vast gap between grade 12 and university standards (Frick, 2008:27). The low skills level in South Africa implies that the economy needs to create low-skill, labour-intensive jobs. This is of course problematic, as China seems to have all but a monopoly in that domain. Structural unemployment in the South African economy implies that in many instances, unemployment can hardly be blamed on the (supposedly unwilling) individual worker, as there simply are no jobs. South Africa's economy is not only characterised by vast proportions of unemployment but also by a lot of *underemployment*. Many, who might be classified as the working poor, earn very low wages, below existing poverty lines. Though there are various measures of poverty, Rogan and Reynolds (2015), based on the most conservative measure, find that 14% of workers live in poor households.

Economic policies have significantly affected employment levels in South Africa. Subsidies to commercial farmers were largely abolished. The agricultural sector shed countless jobs as farmers chose greater mechanisation. Consequently informal settlements – typically the foci of DRAaM – around agricultural towns grew significantly. Similar job loss took place on the mines in the 1990s, as Crush et al. (2006:7) noted, 'which created considerable social disruption and increased poverty in rural [labour] supplier areas'. Under these conditions access to credit is sometimes limited to an asymmetrical relationship between *shebeen* (illegal tavern) owners and community members, as the former take on the role of exclusive credit provider (Barry et al., 2007:176). At the same time, while consumer culture entrenches itself in South Africa, millions suffer the humiliation of being poor in a context where possessions are valued but where they have no realistic opportunity to meet these desires (Ross, 2009:209).

Housing

Basic amenities are often inaccessible in contexts in which millions remain in a perennial state of liminality. It is not merely the permanence of informal structures used to live in that is pertinent to the current discussion. It is also the burden of being an 'illegal' resident of an area and, as mentioned, being trapped in a state of unemployment or underemployment when there is no conceivable alternative and with the accompanying and enduring threat of eviction.

Under apartheid various pieces of legislation were passed to limit black migration into white areas. Yet white South Africa still required black labour in these areas, first as cheap labour and for urban markets and even to raise white children. Migration as such had to be managed as opposed to prohibited. Legislation made it very difficult for black South Africans to settle in urban areas. The state frequently destroyed informal settlements in the 1970s and 1980s. Many communities were forcibly relocated in order to achieve the idealised apartheid urban geography. Today still, informal land occupation is criminalised (Huchzermeyer, 2010:142), adding a further layer to the suffering of millions of South Africans.

As influx control measures were abolished towards the end of apartheid and rural jobs shed, millions of South Africans started to move to urban centres in the hope of finding employment. Yet there is a severe shortage of brick and mortar homes in South Africa. The government was and still is faced with the daunting prospect of providing housing and related services, such as water and sanitation, to millions who did not have access to these amenities before. It is therefore no surprise that millions of South Africans still live in informal structures, which often render them vulnerable to various hazards such as fires and floods. Millions of citizens therefore live without adequate access to water and electricity or without the ability to pay for such services, where these are available. According to the principle of cost recovery, if one cannot pay for services, one does not receive any. An exception is that each household receives a basic amount of free water each month. Some illegally connect themselves to electricity infrastructure, often with fatal consequences.

Housing policy over the past two decades has taken on the form of standardised interventions. Framing an issue such as housing delivery as a technical matter de-politicises it. Thereby this essentially political issue is passed on to managers, who are prone to standardised delivery initiatives, typically in the form of a once-off capital subsidy. Not only does this leave little room for collective reflection on local needs, but as Huchzermeyer (2003:600) argues, it constitutes a form of social control, which fuels feelings of individual entitlement to these standard products.

Such a sense of entitlement only perpetuates the problematic of standardised services.

Local government often attempts to upgrade informal settlements *in situ*, while in other cases residents are indeed forcibly resettled (Huchzermeyer, 2010:139). The former implies an incremental provision of basic amenities in the area of informal occupation. The latter clearly has all manner of potential adverse consequences, not to mention potentially disturbing experiences of déjà vu. Relocation often disrupts social networks, livelihoods, schooling and access to basic municipal services. Such interventions are also ignorant of the socio-political forces which produce housing problems, in other words, why people settle the way they do. Furthermore, the objective of eradicating informal settlements may be inappropriate in many cases. In post-apartheid South Africa a new class of cash-poor households emerged, dependent on backyard shacks as a source of rent income. Unfortunately the practice is often frowned upon by ignorant authorities even though one might easily argue that it is in fact a positive consequence for many new homeowners. They now have a fixed asset to leverage for generating income. The renters have quasi-legal access to a place to stay in relative proximity to work. As Huchzermeyer (2009:265) notes, 'Preventing the use of corrugated iron or so-called "zincs" for backyard letting or informal trade stalls means cutting off a lifeline'. Simply put, there are many logical reasons people settle the way they do.

The housing issue as it pertains to common South African dangers cannot be divorced from other socio-economic and political issues. The post-apartheid model of cost recovery means that many households with access to electricity infrastructure still make use of fuel sources such as coal and paraffin, as they cannot afford electricity. The same applies to water, where pre-paid water meters are used with only a limited monthly allocation of free water. In effect, service delivery has meant that 'freedom' or 'liberation' has in many cases been accompanied by useless infrastructure as an artefact or symbol of continuing oppression pervading the private domains of millions of South Africans. In this sense, housing for many symbolises liminality through insecure tenure, in addition to liminality through incomplete liberation.

Health and hunger

Even though South Africa has, in recent history, been a net exporter of food, the country faces a structural food insecurity problem, which is primarily driven by chronic poverty and unemployment, millions of citizens go hungry (Altman et al., 2009a:348). The Human Sciences Research Council (HSRC) reports that 26.9% of boys and 25.9% of girls 0 to 3 years of age are stunted (short for age), indicative of long-term malnutrition (HSRC, 2013:2). Hunger in South Africa is geographically widespread and occurs in both urban and rural areas. In fact, it seems that the locus of hunger in Southern Africa might be shifting from rural to urban areas, as urbanisation increases and rural-to-urban remittances increase, as opposed to remittances in the opposite direction, previously a major feature of the South African political economy. Poorer households, unlike their wealthier counterparts, often cannot afford a healthy diet. Diets are typically high in fat and carbohydrates but low in protein and micro-nutrients. Moreover, rising food prices disproportionately affect the urban working class and the unemployed, who spend a larger proportion of available funds on food when compared to more affluent segments of society.

As South Africa is a water-stressed country, rain-fed agriculture remains risky, especially in subsistence farming contexts. It is often, however, in such contexts that capital reserves and access to adequate fertile land are limited, in part a result of the degradation emanating from the overcrowding discussed previously. Projected climatic changes could increase the severity of extreme weather events, and as such it will logically exacerbate rural degradation (Aliber and Hart, 2009:436).

Food security also has a gender dimension, further underlining the intersectionalities of affliction for many black women. Policy interventions tend to focus on commercial agriculture and as such largely ignore labour performed by women in subsistence farming contexts, which generally is far more substantial than the labour performed by men (Altman et al., 2009a:357). Women in subsistence farming are, crucially, mostly the ones responsible for feeding and caring for children (HSRC cited in Reddy and Moletsane, 2011:2). Therefore, assistance to (largely women) subsistence farmers might often have many additional and indirect positive consequences.

Any discussion of food insecurity in Southern African ought to heed its connection with HIV/AIDS. HIV/AIDS–related morbidity and mortality can lead to losses of household labour reserves and therefore decreased cash and/or household food production. Extended family networks simply become overwhelmed as relatives take in sick family members or orphans.[2] In some cases, adult mortality eventually manifests as child-headed households, while losses of productive labour may through reduced income and purchasing power lead to so-called erosive coping mechanisms. This disparate category includes instances in which productive assets, such as animals or implements, are sold to buy food and where young women engage in transactional sex in order to afford food, once again potentially facilitating infection. Regardless of the preceding circumstances, whether AIDS–related or not, transactional sex is common in both rural and urban areas.

Gender-based violence and asymmetrical, gendered power relations are important additional causes of HIV infection in South Africa. Women are less likely to request condom use when material gain is at stake (Dunkle et al., 2004:1589). Furthermore, physical and sexual abuse of women leads to alcohol consumption and unprotected sex (Pitpitan et al., 2012).

This subtheme of health as it relates to hunger is further intertwined with various urban policy issues, including the policy of recovering costs for municipal services. The inability to purchase water and electricity is associated with numerous health concerns (Mokwena, 2009). These can, for example, manifest as respiratory disease due to coal burning in winter and gastrointestinal diseases due to a lack of clean water. As wealthier households are typically smaller than poorer households, the 6,000 litres of free water supplied to households is actually skewed in favour of richer households.

The cumulative effects of poverty, hunger and HIV/AIDS and a lack of technological and financial means to extend their lives was exacerbated for much of the Mbeki administration by the state's refusal to provide free access to antiretroviral treatment. According to Johnson (2009:2), some 760,000 eligible adults still had not received treatment by 2008, mostly due to a lack of infrastructure and trained health care workers in many parts of the country. Despite rapid improvement, the situation has not as yet been fully resolved under the Zuma administration. Many South Africans who need anti-retrovirals (ARVs) still cannot access or benefit from these drugs on account of numerous issues bound up in overlapping afflictions such as those discussed in this chapter. Examples in this regard include stigma and insufficient nutrition to tolerate treatment (Daftary and Padayatchi, 2012). The discussion on public health now concludes by briefly focusing on substance abuse as an often neglected though very important matter.

Substance abuse has obvious consequences for physical health. It is also a manifestation of structural violence and, as such, provides evidence of the less-tangible modes of suffering experienced by many South Africans. As with most territories, alcohol abuse is quite common in South Africa. New data in this regard remain rather scant. Parry (2005:426), based on 2000 statistics, nevertheless notes the per-capita consumption of South Africans as 'among the highest in the world'. More recently, the Western Cape especially has seen an overwhelming increase in methamphetamine abuse, or *Tik*, as it is colloquially known. Drugs such as *Tik* are cheap, easily manufactured and therefore quite accessible. While intoxicating substances in general may offer an escape from an unbearable daily life, Ross (2009:28), based on her fieldwork in Chris Nissen Park in Cape Town, also argues that drinking offers a routine escape. In a context in which there is no employment, it provides something to do.

Substance abuse is not limited to any particular segment of society. Still, for many this is an additional layer of affliction, while it is sometimes intertwined with other unfortunate social ills. The impact of substance abuse on households can be severe. Spending the little money available on alcohol or drugs can lead to or exacerbate malnutrition. Furthermore, the high incidence of foetal alcohol syndrome in many parts of the country entails many South Africans suffering their entire life on account of alcohol abuse during pregnancy. The *dop* system has been instrumental in instilling such drinking habits, especially in the Western Cape.

Some engage in illegal and other dangerous activities, such as theft, prostitution and drug sales, in order to survive and to feed their addiction (Hagen, 2008:20). The obvious implication is that when caught and jailed, the individual in question and her family suffer additional trauma and, ironically, additional financial difficulty. Drug abuse, in addition to its impact on the user's body, also potentially threatens the bodily integrity of others through drug-related violence, including violence often associated with organised crime.

Interpersonal violence

South Africa is notorious for being an exceptionally violent country, even in the narrow conception of the term. Many South Africans have been socialised into a context which seemingly continuously recreates a vicious cycle of violence. Interpersonal violence in South Africa takes on various forms. It has historically been associated mostly with Politics (capital 'P'), though it might also be interpreted as having been and still being related to politics with a small 'p'. The struggle against apartheid and the violent oppression of resistance by the police has had an enduring effect on residents. Witnessing police brutality or the necklacing of suspected apartheid state informers has, for example, had an impact on how many South Africans come to view the world.[3] As Marks (2002:20), in writing on the link between social cohesion and HIV/AIDS, notes,

> The scars of such brutal social conflict remain etched on post-apartheid South Africa. The violence that characterised the last days of apartheid has not disappeared, except in its political form. It has proven all the more intractable because long before the last phase of the struggle South Africa had been a very violent society.

Violence often relates to other crimes, such as theft and activities linked to gangs. Many South Africans turn to crime in order to make ends meet. Here it should be noted that despite post-1994 (white) middle-class fears, those most severely affected by crime remain impoverished South Africans. Space does not allow in-depth analysis of all manifestations of interpersonal violence in South Africa. In what follows I have chosen to focus on a particular set of violent acts for which the country is particularly notorious.

South Africa has one of the most heinous histories of sexual violence in terms of both prevalence and the particular nature of these crimes. Much of the violence experienced in South Africa affects women and children, even babies. Violence towards children affects the future, as society continually produces violent adolescents and violent men. It seems that hardly a week goes by without numerous instances of sexual or gender-based violence being reported in the media, while each instance seems to be more unspeakable than the previous. South African sexual violence statistics are the highest in the world (Bhana, 2012:355). It is a society which produces a mind-boggling number of rapists and wife beaters. Many of these cases are not reported, as victims often blame themselves and survivors frequently experience discrimination by the police and the courts (Moffett, 2006:132). Beside the shock and lasting emotional harm, such incidents tend to cause sexual violence, and gender-based violence has practical implications for everyday life. While many women have to bear the trauma of at least one instance of violence in their lives, many others are constantly subjected to such intimate violations.

For many South Africans, as the conception of layers or intersectionalities of suffering suggests, this is yet another affliction they have to bear. I do not profess that there are clear explanations befitting every instance. Given the complexity of diverse social realities, a discussion at the current level of abstraction can only offer potentially useful theories and components thereof. Influx control measures under apartheid, for example, entrenched patriarchal gender relations. The system restricted women's rights to find employment in urban areas, as they were not considered primary breadwinners by the Calvinist Afrikaner-run state. Black women were only granted six-month work permissions in urban areas on condition that they leave their children behind in homeland areas. This would, it was believed, serve as an incentive for these women to return home.

Many women and girls, in the grip of deprivation, turn to prostitution. This can be dangerous, as it often exposes them to violence. Other girls become involved with older men as a means for income. These girls are often exposed to HIV and are caught up in relationships characterised by extremely asymmetrical power relations. The man might provide resources or perhaps even an escape from an otherwise unbearable life. Hence these girls offer a brand of susceptibility ripe for older men to capitalise on. Other forms of persistent gendered intimidation and manipulation include instances in which girls have to manage advances and sexual abuse from teachers, family members and other older men. The implications of gender-based violence are profound. Imagine being afraid of such vast categories as an entire gender and especially those older than you when you are merely a teenager. These are in fact the type of mental scars borne by many women and teenage girls.

Part of the larger point to be made in this chapter implies that human beings have memories. In particular, when you experience such multiple overlapping instances of physical and psychological violence for much of or even your entire life, the psychological effects are surely severe. It is highly unlikely that sexual abuse or even chronic or acute periods of feeling physically insecure as a girl, will not have an enduring effect on women throughout their lives. Indeed, psychologists termed this type of enduring emotional scarring complex post-traumatic stress disorder, though many understandably advise against such a broadening of the concept of PTSD. These authors rather prefer the concept disorders of extreme stress not otherwise specified (DESNOS), which is perhaps also quite unsatisfactory for its lack of specificity (cf. Luxemberg et al., 2001). Nevertheless, an approximate cluster of symptoms, as a result of protracted periods of feeling helpless, especially during the earlier years of life, has been observed. The condition manifests as anxiety, anger and rage brought on by flashbacks and associations of current stimuli with previous trauma, as well as excessively defensive behaviour. Defensive behaviour learned as a means of coping with unwanted stimuli often continues to inhibit social functioning after the initial stimulus has abated. Even worse, though, is that for many, the stimulus does not abate.

In seeking to explain sexual and gender-based violence, we must first distinguish between offering an explanation and presenting excuses. The latter surely has no merit. Furthermore, I do not have the space to offer a comprehensive analysis of the former. The following few points, which have been raised in explaining gender-based and sexual violence as manifested in South Africa, merely intend to serve the general argument pertaining to structural violence as part and parcel of living for many South Africans. Although patriarchy and gender-based and sexual violence occur across racial and class divides, the matter of concern here is with the layers of suffering many South Africans endure. The following points therefore are offered to provide some insight regarding such practices in contexts afflicted by many of the forms of suffering discussed in this chapter. Moffett (2006) argues that sexual violence is fuelled by events of the immediate past. She notes that there is often a marked increase in violence during the initial stages of rebirth after a prolonged period of oppression. Indeed, the first decade of democracy saw a marked increase in sexual violence against women, children and men. She also argues that the *othering* of women is a consequence of the othering practices entrenched in apartheid's inner workings. Men, who experienced othering and hence were once at the whim of a colonial 'master', are now perpetrating a similar deed in order to assert themselves. However, as the analysis in this chapter, interpreted along with Moffett's argument, might suggest, deep-seated forces which fuel such needs for self-assertion, might not have ended with apartheid. Social marginalisation through various forms of deprivation, and by logical extension the accompanying and consequential emotions such as humiliation, helplessness and emasculation, continue. Consider, for example, the potentially emasculating experience within a patriarchal context of being unemployed or underemployed for decades. In such cases violence offers a way of controlling others in the wake of an individual's feelings of powerlessness. The unfortunate victims are those physically weaker than the perpetrator.

This discussion on the horrific trends in sexual and gender-based violence in South Africa as a result of and a constituent part of the broader set of processes defined as structural violence might best be concluded by revisiting Adam Small's play. To reiterate, this type of response to structural violence cannot be condoned in any way. Furthermore, as is the case with *Diekie*, who is sentenced to death by hanging, if they were to find mitigating circumstances for one, they would have to find similar mitigation for thousands, if not millions.

Conclusion

This chapter conceptualised the manifold forms of suffering in contemporary South Africa by drawing on diverse bodies of literature. The concept of structural violence – manifested in both more and less tangible suffering – binds these bodies of literature and as such serves as a counterpoint to contemporary disaster studies epistemologies. Though much has changed in South African society post-1994, much has remained the same as new iteration of older concerns is manifested. Framing matters of danger in the idiom of risk assessment and management is both a symptom of and reflective of the dialectic of South Africa's transition to democracy. Even though many freedoms have been instilled through institutions, such as the Bill of Rights, processes of structural violence prevail. It is the discursive forces enabling structural violence that have changed.

Clearly South Africa's current order, informed by relative conservatism associated with aspects of (neo) liberal political economy, is not working for millions of South Africans. It is painfully evident that most South

Africans do not have much by way of substantive freedoms as conceived of by Amartya Sen (1999). Stated differently, equality of opportunity is only a pipe dream for the average South African, subjected to the various forces constitutive of the broader processes of structural violence. Totalised developmental discourse, such as one-size-fits-all housing solutions or commercialising subsistence farming, merely paints over what remains a largely illegitimate South African political economy.

Assigning risk profiles to the country's many municipalities potentially falls far short of capturing the diverse complex realities pertaining to issues of danger but also, more broadly, the complexity of pressing challenges. The discussions within this chapter suggest that disaster risk assessment and management might not always be the most useful idiom in which to frame a particular population's most pressing concerns. This does not mean that interventions specifically targeted at disaster prevention, mitigation and preparedness have no purpose. However, development should not lose sight of the broader (illegitimate) political economic context. Furthermore, interventions typically based on cultural conceptions of danger cannot be a substitute for the type of macro-level social change promised but not significantly delivered by democratisation. Unfortunately this may be neglected when activities such as DRAaM create the impression that meaningful efforts are being made in alleviating human suffering.

Notes

1 The Gini coefficient is a measure of income inequality. It ranges between 0 (perfect equality) and 1 (absolute inequality).
2 It should, however, be noted that by no means all instances of orphaning are AIDS related (Nduna and Jewkes, 2012:1018). However, Udjo (2011) finds that contemporary increases in orphaning are due to AIDS.
3 Necklacings are a brutal form of public execution, typically used to make a statement 'to make an example of someone' for others to witness. The method entails a rubber tire placed around a person's chest and arms, soaked in petrol, which is then lit. These were common in South Africa during the 1980s and early 1990s to punish suspected government informants. In 2008 this technique was once again employed in some of the country's poorest areas, this time against immigrants from other African states.

References

Aliber, M. and Hart, T. G. B. 2009. Should Subsistence Agriculture be Supported as a Strategy to Address Rural Food Insecurity? *Agrekon*, 48(4): 434–458.
Altman, M., Hart, T. G. B. and Jacobs, P. T. 2009a. Household Food Security Status in South Africa. *Agrekon*, 48(4): 345–361.

Barry, M. B., Dewar, D., Muzondo, I. F. and Whittal, J. F. 2007. Land Conflicts in Informal Settlements: Wallacedene in Cape Town, South Africa. *Urban Forum*, 18(3): 171–189.

Bhana, D. 2012. 'Girls Are Not Free': In and Out of the South African School. *International Journal of Educational Development*, 32(2): 352–358.

Bond, P. 2005. *Elite Transition: From Apartheid to Neoliberalism in South Africa.* 2nd ed. Pietermaritzburg: University of Kwazulu-Natal Press.

Carnegie Commission of Investigation on the Poor White Question in South Africa. 1932. The Poor White Problem in South Africa: Report of the Carnegie Commission. Stellenbosch: Pro ecclesia.

Cock, J. 1980. *Maids and Madams: A Study in the Politics of Exploitation.* Johannesburg: Raven Press.

Crush, J., Frayne, B. and Grant, M. 2006. Linking Migration, HIV/AIDS and Urban Food Security in Southern and Eastern Africa. South African Migration Project (SAMP); International Food Policy Research Institute (IFPRI); The Regional Network on HIV/AIDS, Livelihoods and Food Security (RENEWAL). Online: http://programs.ifpri.org/renewal/pdf/UrbanRural.pdf (Accessed 23 September 2013).

Daftary, A. and Padayatchi, N. 2012. Social Constraints to TB/HIV Healthcare: Accounts from Coinfected Patients in South Africa. *AIDS Care: Psychological and Socio-medical Aspects of AIDS/HIV*, 24(12): 1480–1486.

Degenaar, J. 1980. The Concept of Violence. *Politikon: South African Journal of Political Studies*, 7(1): 14–27.

Dunkle, K. L., Jewkes, R. K., Brown, H. C., Gray, E. G., McIntyre, J. A. and Harlow, S. D. 2004. Gender-based Violence, Relationship Power, and Risk of HIV Infection in Women Attending Antenatal Clinics in South Africa. *The Lancet*, 363(9419): 1415–1421.

Frick, B. L. 2008. The Profile of the Stellenbosch University First-Year Student: Present and Future Trends. Research Report: Centre for Teaching and Learning, Stellenbosch University. May 2008. Stellenbosch.

Galtung, J. 1969. Violence, Peace and Peace Research. *Journal of Peace Research*, 6(3): 167–191.

Gordon, M. B., Roch, S., Depuiset, M. A. and Calemczuk, R. 2012. Are Minorities Over-Represented in Crime? Twenty Years of Data in Isère (France). April 2012. Online: http://hal.archivesouvertes.fr/docs/00/68/57/67/PDF/MinoritiesAndCrime_2012-04-05.pdf (Accessed 24 September 2013).

Gordon, R. 1988. Apartheid's Anthropologists: The Genealogy of Afrikaner Anthropology. *American Ethnologist*, 15(3): 535–553.

Hagen, G. 2008. *The Impact of Methamphetamine (tik) Use on the Workplace.* Master's dissertation. University of Stellenbosch, Stellenbosch.

Huchzermeyer, M. 2003. A Legacy of Control? The Capital Subsidy for Housing, and Informal Settlement Intervention in South Africa. *International Journal of Urban and Regional Research*, 27(3): 591–612.

Huchzermeyer, M. 2009. The Struggle for *In Situ* Upgrading of Informal Settlements: A Reflection on Cases in Gauteng. *Development Southern Africa*, 26(1): 59–73.

Huchzermeyer, M. 2010. Pounding at the Tip of the Iceberg: The Dominant Politics of Informal Settlement Eradication in South Africa. *Politikon*, 37(1): 129–148.

Human Sciences Research Council (HSRC). 2013. Nutritional Status of Children. Media Release No. 2. 6 August. Online: https://www.google.com/search?q=Human+Sciences+Research+Council.+2013.+Nutritional+Status+of+Children.+Media+Release+No.+2.+6+August.+&ie=utf-8&oe=utf-8&client=firefox-b&gfe_rd=cr&ei=6Xc9V9C_NI2p8wfX_LeYCg (Accessed 19 May 2016).

Johnson, L. 2009. *Access to Antiretroviral Treatment in Adults. National Strategic Plan: HIV and STI. 2007–2011*. Cape Town: Children's Institute, University of Cape Town.

Joubert, E. 2001. *Die Swerfjare van Poppie Nongena*. 5th ed. Tafelberg: NB Publishers.

Luxemberg, T., Spinazzola, J., et van der Kolk, B.A. 2001. Complex trauma and disorders of extreme stress (DESNOS) diagnosis, part one: assessment. *Directions in Psychiatry*, 21, 373–392.

Marks, S. 2002. An Epidemic Waiting to Happen? The Spread of HIV/AIDS in South Africa in Social and Historical Perspective. *African Studies*, 61(2): 13–26.

Moffett, H. 2006. These Women, They Force us to Rape Them: Rape as Narrative of Social Control in Post-apartheid South Africa. *Journal of Southern Africa Studies*, 32(1): 129–144.

Mokwena, L. 2009. Municipal Responses to Climate Change in South Africa: The Case of eThekwini, the City of Cape Town, and the City of Johannesburg. Centre for Policy Studies Research Report Number. 113. Online: http://africaclimatesolution.org/features/Municiple_Responses_to_climate_change_in_South_Africa.pdf (Accessed 11 March 2012).

Nduna, M. and Jewkes, R. 2012. Disempowerment and Psychological Distress in the Lives of Young People in the Eastern Cape, South Africa. *Journal of Child and Family Studies*, 21: 1018–1027.

Parry, C. D. H. 2005. South Africa: Alcohol Today. *Addiction*, 100: 426–429.

Pitpitan, E. V., Kalichman, S. C., Eaton, L. A., Sikkema, K. J., Watt, M. H. and Skinner, D. 2012. Gender-based Violence and HIV Sexual Risk Behaviour: Alcohol Use and Mental Health Problems as Mediators among Women in Drinking Venues, Cape Town. *Social Science and Medicine*, 75(8): 1417–1425.

Reddy, V. and Moletsane, R. 2011. The Gendered Dimensions of Food Security in South Africa: A Review of the Literature. Human Sciences Research Council: Policy Briefing. Online: www.hsrc.ac.za/Document-4061.phtml (Accessed 11 March 2012).

Rogan, M. and Reynolds, J. 2015. The Working Poor in South Africa, 1997–2012. Institute of Social and Economic Research (ISER). Working Paper No. 2015/4. Rhodes University: Grahamstown.

Ross, F. C. 2009. *Raw Life, New Hope: Decency, Housing and Everyday Life in a Post-Apartheid Community*. Cape Town: University of Cape Town.

Sen, A. 1999. *Development as Freedom*. Cape Town: Oxford University Press.

Small, A. 1999. *Kanna hy kô hystoe*. 11th ed. Cape Town: Tafelberg Publishers Ltd.

South African Council for Educators (SACE). 2011. Teacher Migration in South Africa: Advice to the Ministries of Basic and Higher Training. June. Online: http://www.sace.org.za/upload/files/TeacherMigrationReport_9June2011.pdf (Accessed 24 September 2013).

Statistics South Africa (StatsSA). 2016. Poverty: Key Statistics. Online: http://www.statssa.gov.za/?page_id=739&id=1 (Accessed 19 May 2016).

Thomas, C. G. 2010. *The Discourse on the Right to Housing in Gauteng Province, 1994–2008*. Master's thesis. University of the Witwatersrand, Johannesburg.

Udjo, E. O. 2011. Magnitudes and Trends in Orphanhood among Younger Persons in the Era of HIV/AIDS in South Africa, 2001–2015. *African Population Studies*, 25(2): 267–285.

Wolpe, H. 1972. Capitalism and Cheap Labour-power in South Africa: From Segregation to Apartheid. *Economy and Society*, 1(4): 425–456.

4 Modes of institutionalisation

Introduction

Emerging discourses such as DRAaM have mirrored the dialectics of South Africa's transition, from one illegitimate political economy based on racialist modes of governing to another founded upon aspects of neoliberalism. This has been evidenced by the country's notorious political history and the relative lack of meaningful change for many South Africans after apartheid. Such an assertion, however, requires additional support in analyses of the particular currently implicated discourse. Institutionalisation has rendered a particular commodification of disaster reduction, founded upon generic operationalisation of DRAaM code and associated technologies. This has rendered various products, packages and promotional strategies designed to meet convenient interpretations of legislative compliance. With commodified expertise there came totalisation with which, in the case of DRAaM, I wish to argue, came the imposition of a foreign ontology. Such imposition is inevitable considering relationship between knowledge and power, as explicated in the previously cited work of Michel Foucault. In addition, I do wish to heed Blaser's (2013) general argument that any analysis already includes within it an idea of how things should be. As such, different accounts of particular phenomena by for example researchers or residents enact different worlds, worlds that are often in conflict with one another, though typically enacted in conjunction with an asymmetrical distribution of agency (Blaser, 2013:552). Hitherto, dominant perspectives have been informed by appropriations of difference under an all-encompassing modernity. Political ontology is concerned with how and why certain worlds are enacted and others not and how to enact more useful worlds. The institutionalisation of DRAaM, a universalising modern discourse, in South Africa has been accompanied by a set of politicised though unarticulated ontological assumptions that translate into ritualised practices with unfavourable implications for knowledge production.

The analysis that follows draws on the conceptual framework outlined in the introduction to this book, though drawing on the notion of *worlding* described in the previous paragraph. It is largely concerned with neoliberal governmentality, including the institutionalisation of discourses into market-able 'expert' forms of knowledge. The broad metanarrative of neoliberalism has shaped and been shaped by the institutionalising of numerous systems of expert knowledge based on the codification of theoretical knowledge. As a result the governed expert consultant meant to offer 'knowledge products', through continued (re)enactment of her professional identity based upon a limited expert code, falls prey to instrumental reason. It is reason in the service of preconceived interpretations of the world and the preservation of institutions and senses of self, associated with a given professional identity. It is also reason potentially closed off to a variety of local *worldings* and associated praxes possibly more appropriate to many of the particularities described in the previous chapter.

The chapter is divided into two sections. The first part outlines the industry by offering a political economy of DRAaM in South Africa. It explains who the main actors are, how they enact DRAaM and how they in the process relate to one another. The second half of the chapter considers the political ontology of the matter. Here I emphasise some problematic assumptions this industry is built on and how DRAaM and its particular forms of instrumental reason produce consequences that are political and undesirable regarding the post-apartheid state-building project. The indus-try's fixation on superficial and generic performance indicators, stemming from international policy convention, has resulted in the institutionalisation of DRAaM in South Africa being largely de-politicising of the substantive issues related to danger. Furthermore, this institutionalisation has simulta-neously been a means towards re-politicising the status quo and the asym-metrical relations between state and society, expert and society and between North and South in general.

Disaster management in South Africa: a political economy

As with most territories, disaster management in South Africa was ini-tially more response oriented. During the 1990s the state initiated sub-stantial initiatives steering policy towards the emerging paradigm of risk management. According to Van Niekerk (2005:109) a major flood on the Cape Flats in 1994 aided this paradigm shift. At the same time a strong local lobby developed which helped to ignite a legislative pro-cess eventually culminating in the Disaster Management Act (the Act) 57 of 2002 (DMA), published on 15 January 2003. The DMA requires risk management structures to be established at all levels of government, as such requiring the 'mainstreaming disaster risk reduction (DRR) into development' described previously.

In South Africa the institutionalisation of DRAaM has taken on the form of instrumental reason, with notably the National Disaster Management Framework (NDMF) being the object of such rationalities. It is the key document through which experts demarcate and justify their professional credence. The document was drafted by a combination of state officials, academics and consultants to provide additional guidance to practitioners on the implementation of the DMA, and as such it outlines sequential processes of risk assessment and risk management. The NDMF includes four key performance areas (KPAs) and three 'enablers', which are deemed necessary for the effective implementation of the KPAs. Each of these seven sections outlines aspects municipalities, provinces and national government should have in place or elements which are strongly recommended. The KPAs are (1) institutional arrangements for DRM; (2) disaster risk assessment; (3) disaster risk management; and (4) disaster response and recovery. The three enablers are (1) information management and communication; (2) education, training, public awareness and research; and (3) funding arrangements for disaster risk management. Unfortunately the *means* outlined in the KPAs and enablers have often been portrayed as *ends*, with significant adverse consequences, simply on account of the corollary inertia. As with the typical characteristics of post-industrial capitalism and contemporary neoliberal government, municipalities tend to outsource various DRAaM tasks to consultants, believed to be proficient in applying the idiom of risk assessment and management to the *problématique* of disaster and danger.

As is the case with any other industry, this one too has its modes of commodification. Commodification is not confined to private-sector consultants. The lines between the public and the private sectors in post-apartheid South Africa have been blurred greatly, as academics, students, academic institutions and numerous hybrids of these increasingly permeable categories manifest. Various consultancies have emerged over the past decade, who together with established enterprises in related fields have attempted to provide the now officially mandated DRAaM expertise. Experts have been found in organisations also offering engineering solutions, construction work, environmental management and emergency management. Other consultancies focus on DRAaM exclusively. These consultants have broad educational backgrounds ranging from no formal training to advanced degrees or diplomas in public management, communication studies, engineering and geography. Other consultants have graduated from short courses or obtained certificates, diplomas and even master's degrees in disaster management/ disaster studies. One participant noted that:

> I think when the disaster management act was announced, they saw here is a thing [opportunity] and then all of a sudden everyone was a disaster management specialist. From engineering firms to lawyers, to people working in construction and new companies that were formed and guys who just finished a diploma and now they are specialists. It looks to me as if everyone climbed out of the woodwork and started to consult.
>
> (Author's translation)

But DRAaM expertise hardly extends far beyond superficial readings of the NDMF. As a set of activities DRAaM is closely associated with 'legislative compliance'. To this end consultants are typically hired to produce disaster risk assessments (DRAs) and disaster risk management plans (DRMPs). For municipalities possessing such reports goes a long way towards generally accepted notions of 'compliance'. However, before that can take place the client requires some sensitisation. A host of organisations offer DRAaM short courses, typically used to 'capacitate' officials in DRAaM jargon. It seems that to have received training essentially means to have in part done your job. It is a popular way of spending a municipal DRM budget. Consider the following observation.

> What we have realised with the training of disaster managers is that they are trained and then they are trained again and then they are trained again and then they go to another institution and get another certificate. But when it boils down to practice, they still do not do risk assessment. How can I say? They still do not have their risk reduction programmes that run on the ground.
>
> (Author's translation)

It is worth noting that some participants expressed their doubts regarding the contribution short courses can make as a tool for transferring knowledge and skills.

> [A]nd you see the people who really aren't keen and have no drive or passion to do it. They are there only because they are going to get a certificate. You know and they now say: 'Oh I am competent'. Actually they are not.
>
> (Author's translation)

Other indicators of legislative compliance are met through consultative meetings, which assist in setting up mandated municipal institutions and often even to 'research' aspects on which DRAs and DRMPs are based. In other cases DRAs may be based on secondary analyses of census data or inferences drawn from land-use or land-cover maps, using geographical information systems (GIS). This form of research based consultancy apparently requires very little or in many cases no field research at all. Where fieldwork takes place the process has been hamstrung by the logic of time equals money. The result has been standardised, and superficial knowledge (cf. Van Riet and Van Niekerk, 2012). Buys (2005:4), in elaborating on the NDMF, holds that risk assessment, the subject matter of KPA 2, should be a continuous process in order to ensure that interventions have been effective. In reality risk assessments are conducted by consultants every few years. Regarding DRAaM institutions, KPA 1 of the NDMF (South Africa, 2005:4–24) states that specific 'institutional arrangements' are required for DRM to be truly a 'multidisciplinary' and 'integrated' activity, where DRM is 'mainstreamed into development'. It furthermore requires that provinces and metropolitan and district municipalities, create disaster management centres (DMCs) with permanent staff. According to the NDMF (South Africa, 2005:36), mainstreaming is best achieved through an interdepartmental committee, a structure often invoked as part of a consultancy project. The committee is typically chaired by the head of the DMC. The committee meets every quarter and should include senior members of all municipal departments who can collectively make decisions regarding risk management in their jurisdiction.

The very notion of 'legislative compliance' is malleable, as each consultancy seems to have a slightly different interpretation of what constitutes compliance. Some, for example, make use of 'technical advisory committees' for risk assessment, but others do not.[1] Some make use of community-based disaster risk assessment' (CBDRA) while others do not. These are pliable concepts, which, due to their inevitable association with notions of 'the good' and because they are based on the performance of a professional vocabulary the client is not necessarily privy to, can easily take on the form of *weasel words* (see Roosevelt, 1916), deployed to secure contracts.[2] This profession is governed with the aid of policy documents and DRAaM conceptual code. While there is significant literature on each component, the code in question is often reduced to the $R = H \times V/C$ equation explained previously. The assumption seems to be that this code and the technical measures (the means) outlined in the NDMF offer an optic supposedly applicable across time and space. In a competitive industry where time equals money it is therefore not in-depth and contextualised knowledge of a given locale that these workers specialise in. It is rather the opposite in both instances.

Disaster risk management logically follows from DRA. Key Performance Area 3 of the NDMF (South Africa, 2005:39–53) includes guidance on how to draft policy frameworks and plans and how to incorporate DRM into other on-going programmes. Accordingly DRAs must inform DRMPS, which in turn should inform the municipal integrated development plan (IDP). The IDP is a municipality's principle strategic planning document, which informs decision making on development priorities for a period of five years. Outsourcing DRAaM to consultants who market a basic code and the associated standardised tools to do essentially the bare minimum has been an abysmal failure. Various consultants interviewed noted that municipalities typically attach the DRMP to the IDP as an addendum and claim to have aligned the two and therefore to have mainstreamed DRR into development. Tellingly, no consultant reported that they assist municipalities with this integration.

The infusion of new public management (NPM) into the post-apartheid public service evidenced here manifested in a particular institutional climate, which is partially a remnant of the past. According to Cameron (2009:912) NPM is a set of ideas which includes two sub-categories. The first is the use of private management ideas including more responsive and efficient services, performance agreements and greater autonomy and flexibility for managers. The second sub-category has been more significant in the South African context. It dictates greater use of market mechanisms, such as privatisation and public–private partnerships in service provision.

McLennan and Munslow (2009:275) note that as far as South Africa's service delivery machinery is concerned, bureaucratic hierarchy still dominates as it did under apartheid. Many participants confirmed the associated unwillingness of the bureaucracy to make independent decisions, while the addition of NPM and market mechanisms in particular dictates a broad acceptance that it is often consultants who do DRAaM as opposed to bureaucrats. Consider the following example:

> [A]nd many people are totally . . . want to be fed with a spoon and do not want to think for themselves. He thinks you should think for him. So it is almost a lack of ownership, I would say, for their own process and for their own products, where they must deliver. It becomes the consultant's plan, not the municipality's plan.
>
> (Author's translation)

Therefore 'the way' we do DRAaM in South Africa is mostly top down. The associated lack of independent and creative decision making by lower- to mid-level officials is punctuated by a celebration of neoliberal affinities for 'expert knowledge' and the governing frames of reference directed by very limited notions of 'compliance'. It seems DRAaM in many cases is not much more than a budget to be spent on experts who profess to know how to achieve these conveniently defined objectives. Consider the following quotation: 'The plans do not address the real problems. The plans follow the guidelines of the legislation. It [the plan] is superficial' (Author's translation).

By actively embracing post-industrial capitalism, the fundamentally political questions regarding who is endangered, when and how and conversely who is not are reframed as technical matters to be 'managed' by technocrats. The aforementioned limited interpretations of legislative compliance that rely on consultants to do what is perceived as the entirety of disaster reduction have unsurprisingly led to limited projects initiated under the guise of or in consideration of disaster reduction. The following quotation provides evidence of this point: 'Much of the risk reduction stuff which comes out of the assessments is never operationalised. It is never stated how we are going to implement it' (Author's translation).

There are numerous cases of these reports merely gathering dust, to briefly revert to description by cliché, and where structures such as advisory forums and interdepartmental committees suffer rapid decline. Still the institutionalisation of DRAaM in South Africa has led to increases in staff expenditure and regular consultancy projects running into hundreds of thousands and in some cases even millions of rand. In light of what seems to be very little return on investment, we have to ask: is this how we should be spending these scarce resources? Furthermore, to reiterate a conclusion from the previous chapter, are disasters a priority concern in each and every context?

Closely related to DRAaM's execution is the supposed problem of 'policy implementation', a discourse which has largely characterised analyses of the post-apartheid civil service. Accordingly South Africa boasts supposedly good legislation but an inadequate civil service, incapable of implementing this legislation (Draper et al., 2014; McLennan and Orkin, 2009:1042). Cameron (cited in Cameron and Thornhill, 2009:901) notes that many supposed issues of poor policy implementation in South Africa are instead matters of policy itself. Often policy makers are overambitious in their objectives, in particular the demands these place on state structures. I believe this to be true of South Africa's DMA and NDMF. These much-celebrated documents are attempts at mainstreaming DRR into development. This has meant an enormously cumbersome set of human resource and financial demands, not to mention an implicit call for standardisation and universal precedence of DRAaM above alternative definitions of reality. Consequently mainstreaming DRR into development has largely failed.

As previously mentioned municipalities often form DMCs, staffed by 'disaster risk managers', to conduct multi-sectoral DRM across all municipal departments. But how does one person or department contend with such complex socio-environmental dynamics, even if only through fulfilling a 'coordinating role'? As the previous chapter indicated, isolating a set of causes for one particular or a particular set of afflictions by drawing on one rudimentary code seems rather absurd. Even some of those involved in the field find its breadth challenging. Consider the following quotations: 'I think that because disaster management is so broad and they have so many people to deal with, that makes it difficult'; 'Disaster management can be made out to be as vast as the Lord's grace' (Author's translations).

'Mainstreamed' disaster risk management seems to be an inordinate intellectual pursuit even long before any action can follow. At the same time the prevailing legislation and the requisite enactment of professional identities demands that these managers force complex dynamics and often multiple overlapping afflictions many South Africans face into a very limited conceptual framework.

Truly taking risk reduction to its logical conclusion will necessarily imply a change in the very (dis)functioning of the broad political economy. Instead, however, a conservative industry of generalists has manifested, at best professing to be progressive in the process of establishing an emerging professional identity. Disaster risk assessment and management is conservative both on account of logical tendencies toward inaction and through constitutive vested interests, reliant upon the neoliberalism-infused status quo and associated institutionalised professional identities. It therefore constitutes an industry conflicted at its very core.

Making higher education 'relevant': a political ontology

Higher education institutions have experienced a decline in state subsidy since the end of apartheid. Public institutions have increasingly sourced additional forms of income or third-stream (in addition to student fees and state subsidy earned through conventional teaching and research) income. 'Third money stream' is a very broad category, including diverse sources of funding such as research grants from private donors and public donors, short courses and other consultancy-type activities. Some commentators have voiced concerns about research agendas in some South African contexts that are now determined by those buying their research (Desai and Bohmke, 1996). In this context of declining state subsidy, neoliberalism as a mode of governing and surveillance on university staff often demands the quantitative measurement of outputs, and in the process there has been a shift in power from faculty to administration (Habib et al., 2008:147). In South Africa there have been significant retrenchments and even departmental closures since 1994, as the market has been an increasingly prominent rationality behind university management (Badat, 2009:458, 465; Ensor, 2004:350).

Since the 1990s there has been what Jansen (2004:308) argues is 'an increasing sense among the public that academic offerings should provide vocational training'. The National Department of Education and thereafter, subsequent to restructuring, the Department of Higher Education in particular have reinforced such perceptions and actions. Ensor (2004:351), in reference to the National Qualifications Framework (NQF), notes that

> What entered higher education as a definition of a programme came to be associated with a number of key features: portability, relevance (or responsiveness), coherence and interdisciplinarity (which was offered as the vehicle to achieve relevance).

The term 'interdisciplinarity' referred to in this quotation has often manifested as courses in fields such as DRAaM, developed in a quest for financially beneficial 'relevance'.[3] As Mamdani (2007:98) argues in the case of Uganda's Makerere University subsequent to neoliberal reforms, problem-based curricula commonly labelled 'interdisciplinary' have suffered from severe theoretical impoverishment. In South Africa many DRAaM degree offerings tend to rely largely on service courses, while specialist courses often do not go much further than to outline the DRAaM code described earlier. The possible exception may be one instance where disaster risk has been situated in the environmental and geographical sciences (EGS), which is arguably a discipline or set of disciplines better suited than most for dealing with the disaster problem on account of its simultaneous internal coherence and broad conceptual reach. To be fair, drawing on different bodies of literature from various disciplines or discursive spaces between disciplines inherently sounds like a good idea. But such an immediate positive judgement can be misleading, in particular where it has led to theoretical impoverishment and therefore questionable interpretive acumen.

This lack of interpretive acumen is observable in much of the master's, doctoral and other academic research undertaken in the idiom of DRAaM in South Africa. These studies are often characterised by very limited theory building. In many cases researchers attempt to measure legislative compliance or compliance with international policy conventions by drawing on imprecise indicators based on policy documents. A particular policy document, typically the NDMF, is taken as the solitary benchmark, supposedly above critique and supposedly comprehensive in dealing with the complexities pertaining to disasters in the study area. Many researchers see little need to draw on and speak to existing scholarship. Instead of the anticipated positive results of interdisciplinarity/multidisciplinarity, we have found the reification of policy in the name of 'relevance'. This is of course preposterous, on account of the inherent tension between such universalism and any reasonable denotation of relevance across time and space. As with consultancy projects, training 'disaster risk managers' seems to be a case of predefining the problem(s) before even setting foot in a given context in the few instances where field research actually does take place. If 'relevance' is only achievable through instrumentalised and theoretically impoverished teaching and research, then we have a major problem. Are we to unconditionally adopt Northern discourses and simply research 'how to' make them fit in our context? Such an approach to higher education seems a lot like Bantu education on a global scale, as the South African higher education fraternity by implication is suggesting that 'different types' should be researching and teaching different forms of knowledge. Defining reality is solely the privilege of the dominant North. The Global South should work within the specific set of 'relevant' intellectual styles assigned to it. The problem is exacerbated when the discourse in question is further distilled and standardised in aid of quantitative output.

Many universities have started to engage in DRAaM consultancy and offer short courses in this field, while the same lack of in-depth and critical reflection found in short courses also characterises formal degree and diploma courses. According to Dahlström (2008:8), neoliberal inclinations 'are transforming universities into business like enterprises'. Diploma or degree programmes are developed to cater for the needs of the market, of paying 'clients' or 'consumers' who need to be trained in currently favoured totalised discourses. Furthermore, it is the market, for the most part, that gives substance to this notion of 'relevance'. One can of course argue that terms such as 'the market', 'client' and 'consumer', used to justify the instrumentalisation of higher education, refer to a set of actors who are themselves the products of and governed by some of the very same actors constituting and governing micro and macro-discourses. Thus, higher education institutions have contributed significantly, through tailoring their course offerings and third-money-stream initiatives in shaping and sustaining markets such as the DRAaM industry. In a Global South context, leveraging development discourses, with their accompanying potential for employment, is one way to secure a clientele. However, what if the market is wrong, in a milieu where questioning and debate is dismissed for not being 'relevant'? Vale (2011:22) in writing on the state of the humanities in South Africa, notes that these disciplines (and I would argue theoretically rich education in general) are necessarily relevant. They continuously asks fundamental and reflexive questions, pertaining to more or less appropriate conceptualisations of pressing concerns and even regarding what the most important concerns are or should be. They have inherent value, as they have the potential to enact diverse worlds and praxes, whereas instrumental analyses do exactly the opposite by starting at the general conclusion and then merely making readily accessible pieces of data fit. Regarding the broader sociology of contemporary higher education, the fact that we are training potential (disaster risk) *managers* is significant.

Two interrelated features of the contemporary South African higher educational sector are important in explaining the general context that allows management courses to flourish. First the very field, management, is something South Africans have increasingly held in high regard since 1994 (Jansen, 2004:308). Furthermore, public management, the post-1994 iteration of public administration/native administration, is a particularly fascinating case. The post-apartheid proliferation in management qualifications, in part relates to the proliferation of private higher education institutions, including competition from foreign universities settling in South Africa since the 1990s (Badat, 2009:460). Private higher education institutions are not necessarily a threat to public institutions in terms of direct competition. They are typically less research and more vocationally focused, and they have limited student numbers (CHE, 2009:11). Private institutions have, however, been significant for their part in shaping the general interplay between higher education, state policies and market demands in the post-apartheid commercialisation of higher education. As such, Nayyar (cited in Badat, 2009:458) suggests that they have helped shape a milieu governed by a notion of 'education as business'. The perception of management degrees, which unlock a bright future, does not present the full picture yet. To do this we must link the appeal of such applied qualifications for those South Africans preoccupied with 'preparation for the job market' or, as Jansen (2004:308) puts it, since the late 1990s there has been 'a sense among the public that more vocationally-oriented courses are needed'. Some typically historically disadvantaged students view these qualifications as a means toward long-overdue upward social mobility. Many white students view these qualifications as a defence against the perceived threat posed by black economic empowerment (BEE). As I will explain, these students are mistaken, as they conflate instrumentality with employability.

With regard to public management in particular, a new market opened up in the early 1990s, exploited by public and private institutions alike. As the civil service needed to transform, so too did the composition of public administration graduates. This required more 'capacity-building' initiatives in response to the needs of various state structures. Furthermore, many officials want to improve their career prospects by acquiring an additional qualification while working (McLennan, 2007:6). Democratisation therefore provided a critical opportunity for many residential universities to diversify by providing 'relevant' qualifications to this largely untapped market. This required greater coordination of the institution's schedule with that of the new client. As a result, the 'semi-distance learning' teaching model, including a few weeks of contact between staff and students per year, has become increasingly prevalent. On account of the discussion thus far one could argue that post-apartheid discourses of expert knowledge, higher education, public administration, management and relevance converged into an intertwined complex of neoliberalism-infused instrumentalisation.

Indeed, discourses such as DRAaM are found where many of these debates intersect, and as such it has inherited much of the challenges historically faced by public administration, both the discipline and the activity. There is certainly much to be said about the instrumentalisation of many other academic disciplines in post-apartheid South Africa. Still, few are more significantly implicated than public administration. In the South African case DRAaM has largely been built upon the foundations provided by this discipline. The fact that many of the key figures in DRAaM were schooled in public administration likely shaped the particular management-oriented and legislation-reifying approach described in this chapter. The result has been a bizarre process of instrumentalisation (DRAaM) layered upon more instrumentalisation (public administration), taking neoliberal principles of expert consultation and policy documents as its instruments.

Public administration/native administration was the vehicle through which many apartheid bureaucrats were trained. As McLennan and Munslow (2009), cited earlier, explain, it produced graduates who were obedient, following protocol to the letter. The accompanying scholarship had similar 'how to?' as opposed to 'why?' inclinations. There have been a number of reflections on the state of the discipline in terms of both teaching and research. Apparently much of the same criticisms uttered in the early 1990s are still voiced today. Then the most prominent discussions were on the so-called New Public Administration Initiative (NPAI), mobilised by some within the field to rejuvenate it in a more appropriate post-apartheid form. Cameron and Milne (2009:385) list various issues which at that point needed to be resolved. Two issues stand out because they evidently have not been meaningfully resolved. According to these authors, public administration

- was too descriptive: lacking sufficient analytical, explanatory and predictive techniques;
- is reductionist: restricting and reifying public administration to one view of the administrative processes only.

For the vast majority, re-energising the field meant training interventions in order to transform the civil service as well as instilling an explicitly developmental focus in the public service. Many universities would leverage this developmental focus as a tool for revenue supplementation. How to preserve white dominance was replaced with how to do integrated development planning, often in the narrowest, formal and technical sense, often by paying consultants to draft these plans. Thus Cameron and Milne (2009:393) state that the discipline 'has not moved beyond a technical, how to approach'. The discipline has all but severed ties with the social sciences by adopting a near-exclusively managerial approach to serving the public. This persistent technocracy, according to Sindale (2004:671), ignores the very fundamental public administration principle, namely that it is time, place and culture bound. His argument is echoed by Mubangizi and Theron's (2011) who analyse South African business administration curricula. They find that these are largely 'technocratic' as opposed to 'democratic' in nature, as the humanities remain significantly under-represented. The 'democratic' curricula they advocate would essentially focus on instilling the value of participation by the public in decision making. Pithouse's (2009) writing on housing provision argues similarly, suggesting that there is an unfortunate tendency by the state to lead with policy as opposed to politics. The result is that policy is rather forced onto the public instead of following from participation or being sufficiently open ended to make it amenable to ongoing dialogue with the public, diverse ontological statements, worldings and praxes.

Policy fetishism is evident in certificate, diploma, undergraduate and postgraduate degrees offered under the rubric of DRAaM. Students can obtain certificates (one year), diplomas (three years), degrees and master's degrees with specialisation in DRAaM. Public management is often a major or a core course. 'Management' has even featured in the names of degree programmes. Programmes are often offered through the semi-distance learning model. In many cases content closely follows prominent policy documents. Consider the following course titles, clearly linked to the NDMF: Institutional Capacity (KPA 1); Risk Assessment and Monitoring (KPA 2); Risk Reduction Planning and Monitoring (KPA 3); Operation Response and Recovery (KPA 4); Education Training and Awareness (Enabler 2). Sometimes study material covers an immense breadth of work, though unfortunately with very little depth. In one instance a course developer had ambitions of fitting disaster early-warning systems, risk assessment, project management and community-based risk reduction into a single second-year semester module. In some cases sections in the study material are extremely brief, extended by an inordinate number of diagrams and figures copied out of international or national policy documents or field manuals.

Frequently lecturers and study material authors (not always the same person) are not academics, which exacerbates the largely 'how-to-do' nature of such offerings. It is therefore not surprising that reading material often draws heavily on international and local policy documents. Besides the likes of the NDMF, UN documents and international conventions are cited repeatedly, as if they were academic texts, while proclaiming one single truth now and forever. Policy documents and therefore much of the prescribed reading treats the social as static and reducible to simple conceptions of reality. This is unavoidable when lists are literally copied and pasted from policy documents without accompanying explanation and justification, let alone critical reflection. Moreover, if we supposedly want to deliver graduates who are 'work ready', then how are we supposed to teach students, who are more than likely future report writers, writing skills when the example set by the study material is one of ineptitude?

The intertwined relationship between higher education and practice set around a static view of reality presents a major challenge to the skills set of future disaster management professionals. One director responsible for managing a certificate programme told me that their institution's study material is regularly evaluated by industry in order to assess 'relevance' and to ensure that they continue to produce what the market requires. This reveals a dangerous self-referential cycle, where technical training, based on current policy, produces those who assess future training by those same standards. The assumption seems to be that current policy does work and is 'best practice' now and forever, everywhere.

It seems especially regarding theoretical depth that supposed interdisciplinary or multidisciplinary DRAaM programmes are found wanting. My argument is not that practical experience is inferior to theoretical knowledge. It is that if we are going to teach students 'how to do', often using practitioners as lecturers, then education has been rendered all but immaterial. It has become a means to something other than an education (e.g. promotion; professional identity), and in the process theoretical knowledge and professional capacity have been subordinated to time- and space-bound practical knowledge purported to be universal by a small group of academics and practitioners. Based on Drucker's (2001:326) argument that the post-industrial economy requires life-long learners likely the graduates of a theoretically rich education, one might more feasibly suggest that graduates educated accordingly will be better disaster managers. They will be better placed to recognise when DRAaM is and when it is not the most useful way of framing a particular issue. They will have more substantial conceptual tools and a superior ability to continually acquire new, more appropriate conceptual tools with which to understand and define the particular realities they are paid to attend to. Exposure to what Mubangizi and

Theron refer to as democratic curricula may furthermore instil an ethic acknowledging field research as the most appropriate means of gaining insight into these particularities and the means by which appropriate worlds may be enacted. This may produce better knowledge, while the associated dialogue could serve as a resource in aid of a more productive relationship between state and society. These graduates, who would have learned how to absorb a lot of knowledge relatively quickly, would likely in any event only require a few months, if not weeks, to acquire the same degree of familiarity with the DRAaM jargon on which contemporary notions of 'expertise' are based.

Equating or even preferring 'experience' to conceptual analytical skills is to continually reproduce the self-referential cycle between practice and academia mentioned earlier. It is not possible for both to contribute what they are best equipped to do when higher education is continually morphing towards the habits of practice, even sacrificing truly independent scrutiny for a sheep-like obedience to the truths proclaimed by policy makers and 'the market'. This does not produce a complementary relationship between academia and practice. Rather, it impoverishes both as static groupthink replaces critical and more socially inclusive dialogue.

Conclusion

The institutionalisation of DRAaM in South Africa has produced an edifice of expert report writing, short courses and undergraduate and postgraduate degrees where elementary DRR code and policy documents have been deployed similar to scholarly literature in response to policy and its apparent call for professionalisation. Many consider this to be impressive, as industry has 'progressed' away from a focus on disaster response and teaching and research has become more 'relevant'. Unfortunately notions of relevance based on a singular and superficially enacted professional code are untenable. Such instrumental logic is premised upon ahistorical conceptualisations of knowledge and an *atheoretical* understanding of knowledge production manifested in an incestuous relationship between the market and higher education. This presents a particular politics of ontology, which can be explained by stating two misguided assumptions upon which this logic rests.

First, there is the assumption that for higher education to be relevant necessitates instrumentalisation. This is where readily available policy discourses are handy to both higher education and private-sector consultants. They offer something marketable as a service to paying clients, be it through consultancy services or education. Unfortunately the result has been technocracy and superficial training and research, governed by the instrumentalised discourse in question and the whims of the market. Graduates and practitioners only ask how, as opposed to why DRAaM? More well-rounded graduates will most likely have better faculties for recognising the need for a particular conceptual approach and adapting to changes in dominant conceptual approaches. If their education included a strong focus on democratic values in the sense previously explained, they will more likely recognise the need for popular participation, currently all but ignored by many working in DRAaM and its potential for enacting contextually appropriate worlds and praxes.

There is a second corollary assumption pertaining to who has the right to ontological statements and who does not. In this instance it manifests as the assumption that everyone is more or less 'at risk', just as they had previously been disease ridden, 'uncivilised' and 'underdeveloped' in the terms set out in a foreign narrative. The result is a threat to critical and reflexive scholarship and a continuation of the narratives of difference. The North defines the supposed singular modern world, while the South is disciplined through discourses of relevance, in actual fact a by-product of the conceivable range of strategies for professional and institutional survival in a given time and space.

Notes

1 A technical advisory committee is a team of specialists usually co-opted from municipal departments to contribute to the risk assessment directed by a consultant. It is essentially a means of integrating into the risk assessment process the requisite specialist skills not offered by the service provider. This does not appear to be a very common practice.
2 By 'weasel word', I mean words at first glance associated with the 'good' but upon closer inspection decidedly lacking in clarity. It is an insubstantial rhetorical device.
3 'Multidisciplinarity' refers to scholarship in which various disciplines apply their particular perspective to a topic. This may be compared to transdisciplinarity, which assumes that answers to research problems lie outside of disciplinary knowledge. This approached draws on various perspectives, not necessarily explicitly acknowledging a disciplinary contribution to that understanding (see Goebel et al., 2010:478).

References

Badat, S. 2009. Theorising Institutional Change: Post-1994 South African Higher Education. *Studies in Higher Education*, 34(4): 455–467.
Bankoff, G. 2001. Rendering the World Unsafe: 'Vulnerability' as Western Discourse. *Disasters*, 25(1): 19–35.
Bell, D. 1973. *The Coming Post-Industrial Society*. London: Heinemann.

Blaser, M. 2013. Ontological Conflicts and the Stories of Peoples in Spite of Europe: Toward a Conversation on Political Ontology. *Current Anthropology*, 54(5): 547–568.

Buys, L. J. 2005. Package of Information to National Platforms for Disaster Reduction. Department of Provincial and Local Government. Online: http://www.unisdr.org/2005/mdgs-drr/national-reports/South-Africa-report.pdf (Accessed 3 May 2013).

Cameron, R. 2009. New Public Management Reforms in the South African Public Service: 1999–2009. *Journal of Public Administration*, 44(4.1): 910–942.

Cameron, R. and Milne, C. 2009. Minnowbrook, Mount Grace and the State of the Discipline. *Journal of Public Administration*, 44(2): 380–395.

Cameron, R. and Thornhill, C. 2009. Editorial – Public Service Reform in South Africa: 1999–2009. *Journal of Public Administration*, 44(4.1): 897–909.

Council on Higher Education (CHE). 2009. The State of Higher Education in South Africa: A Report of the CHE Advice and Monitoring Directorate. Pretoria.

Dahlström, L. 2008. Critical Intellectual Work: An Endangered Tradition under Neoliberal Regimes. Paper prepared for Meeting Global Challenges in Research Cooperation: A human rights perspective. Uppsala, Sweden, 27–29 May 2008.

Desai, A. and Bohmke, H. 1996. Death of the Intellectual, Birth of the Salesman. *Debate*, 3: 10–34.

Draper, A., Draper, C. and Bresick, G. 2014. Alignment between Chronic Disease Policy and Practice: Case Study at a Primary Care Facility. *PLoS ONE*, 9(8): 1–8.

Drucker, P. F. 2001. *The Essential Drucker*. New York: Collins.

Ensor, P. 2004. Contesting Discourses in Higher Education Curriculum Restructuring in South Africa. *Higher Education*, 48(1): 339–359.

Goebel, A., Hill, T., Fincam, R. and Lawhon, M. 2010. Transdisciplinarity in Urban South Africa. *Future*, 42: 475–483.

Habib, A., Morrow, S. and Bentley, K. 2008. Academic Freedom, Institutional Autonomy and the Corporatised University in Contemporary South Africa. *Social Dynamics*, 34(2): 140–155.

Jansen, J. D. 2004. Chapter 11: Changes and Continuities in South Africa's Higher Education System, 1994 to 2004. In Chrisholm, L. (ed.) *Education and Social Change in Post-apartheid South Africa*. Pretoria: Human Sciences Research Council. pp. 293–314.

Jansen, J. D. and Vale, P. 2011. *Consensus Study on the State of the Humanities in South Africa: Status, Prospects and Strategies*. Pretoria: Academy of Science of South Africa (ASSAf).

Mamdani, M. 2007. *Scholars in the Marketplace: The Dilemmas of Neo-Liberal Reform at Makerere University 1989–2005*. Pretoria: HSRC Press.

McLennan, A. 2007. The Academic/Practitioner Interface in Public Administration in South Africa. Paper presented at the 4th Public Management Conversation, Building the Academic-Practitioner Interface in South African Public Administration. 18–19 April, Lord Charles Hotel, Somerset West.

McLennan, A. and Munslow, B. 2009. *The Politics of Service Delivery*. Johannesburg: Wits University Press.

McLennan, A. and Orkin, M. 2009. 'That Elusive Value': Institutionalising Management Development for Developmental Government. *Journal of Public Administration*, 44(4.1): 1027–1045.

Mubangizi, B. C. and Theron, F. 2011. Inculcating Public Leadership for Citizen Value: Reflecting on Public Administration Curricula. *Administracio Publica*, 19(1): 33–50.

Pithouse, R. 2009. A Progressive Policy without Progressive Politics: Lessons from the Failure to Implement 'Breaking New Ground'. *Town and Regional Planning*, 54: 1–14.

Roosevelt, T. 1916. The Weasel Words of Mr Wilson. Morning Speech by Theodore Roosevelt at 31 May, St. Louis. Online: http://www.theodore-roosevelt.com/images/research/txtspeeches/673.pdf (Accessed 7 February 2014).

Sindale, A. M. 2004. Public Administration Versus Public Management: Parallels, Divergences, Convergences and Who Benefits? *International Review of Administrative Sciences*, 70(4): 665–672.

South Africa. 1998. *The Skills Development Act. Act 97 of 1998*. Pretoria: Government Printer.

South Africa. 2002. *Disaster Management Act. Act 57 of 2002*. Pretoria: Government Printer.

South Africa. 2005. *National Disaster Management Framework*. Pretoria: Government Printer.

Vale, P. 2011. Instead of a Defence: Thoughts on the Humanities at Home and Abroad. *Theoria*, 58(128): 21–39.

Van Niekerk, D. 2005. *A Comprehensive Framework for Multi-sphere Disaster Risk Reduction in South Africa*. Doctoral dissertation. North-West University, Potchefstroom.

Van Riet, G. 2015. *Instrumental Reason and Neoliberal Governmentality: A Critical Analysis of Disaster Risk Assessment and Management in South Africa*. DLitt et Phil thesis. University of Johannesburg. , Auckland Park.

Van Riet, G. and Van Niekerk, D. 2012. Capacity Development for Participatory Disaster Risk Assessment. *Environmental Hazards*, 11(3): 213–225.

5 Knowledge production

Introduction

The current chapter focuses on DRAaM knowledge production in South Africa since the promulgation of the DMA in early 2003. In the process it engages with various governing micro-discourses constitutive of the broader DRAaM edifice and associated professional identities. The chapter is conveniently divided into two sections. The first part discusses academic knowledge production. The second part analyses consultants' reports. However, for reasons outlined in the previous chapter, the two categories, namely 'academic' and 'practice', should rather be viewed as ideal types in the Weberian sense, enacted here to give structure to the analysis.[1]

At this point, one matter needs clarification. Much of the research, and potentially even some of the most theoretically rich research, might pertain to DRAaM only by implication. It has DRAaM application, but it is not framed in that vernacular. The discussion on academic DRAaM knowledge production presented here only extends to those instances in which the research is overtly framed in terms of DRAaM – in other words, where the particular vernacular associated with DRAaM is explicitly employed or where titles, keywords and abstracts evidently include words such as 'DRR', 'DRM', 'disaster risk' or 'risk'. Extending the analysis more broadly would have defied the study's parameters. The point of departure is simply this: there is a growing group of academics and students who are focused on DRAaM, as evidenced by the defining characteristic code referenced in numerous instances previously. It is on this knowledge production that the chapter focuses.

Building on the genealogical analysis employed in Chapter 1, a second Foucauldian research strategy is deployed. The analysis in what follows therefore demonstrates that 'what power does' is to serve instrumental needs as each of DRAaM's pivotal micro-discourses eventually falls prey to the bureaucratising, de-contextualising, de-historicising and de-theorising habits of instrumental reason and neoliberal governmentality. Academic and especially consultancy-based knowledge production tends to redefine the fundamentally political issue(s) of human suffering side by side with rela- tive *and* absolute opulence in the simplistic code of the industry. This act of framing a hydra with structural causes in technical terms is de-politicising and re-politicising. A pressing complex set of political issues is recast as a single technical issue, amenable to expert intervention and incrementally (if at all) managed change, as if these everyday afflictions are not that urgent. Re-politicisation relates to the relationship between experts who are privy to the code in question and the stated beneficiaries who become additional objects of technical probing, manipulation and as such violation, inevitably to first and foremost pay professionals' and their employers' salaries.

Academic knowledge

South African academics have approached the DRAaM problematic from various disciplinary, multidisciplinary, interdisciplinary and transdisci- plinary stations (hereafter multidisciplinary). Examples of such disciplin- ary bases include public administration, sociology, communication studies and environmental and geographical sciences (EGS). The multidisciplinary contexts referred to here are mostly research and consultancy centres and postgraduate programmes, with the exception of RADAR/DiMP, which has for the most part maintained a particular disciplinary (EGS) footing. The following discussions focus on two interlinking themes pertaining to structuring, governmental power and instrumental reason in South African DRAaM scholarship, namely, multidisciplinarity and the use of legisla- tion in analyses. Both of these themes are manifestations of a more gen- eral trend, namely a general dearth of meaningful conceptual analysis in DRAaM–related research outputs. With regards to both themes, the discus- sion highlights certain instrumentalist trends. Academic research falls into very similar traps to consultancy, therefore reinforcing the coalescent, self- referential relationship with the private sector, discussed previously.

Multidisciplinarity

Many analyses of DRAaM in South Africa would likely be termed scholarly by most, even where such a concept is very narrowly defined. Many though not all of these studies draw on disciplinary theory or theory from fields of study related to or relatable to DRAaM, such as food security (Kasie, 2009) and systems theory (Coetzee, 2010). Mgquba and Vogel (2004) offer an anal- ysis informed by but adding to the PAR Model. This is similar to authors such as Mavengere (2011), who use conceptual frameworks typically associated with DRR/DRM, though adding more substance through more established disciplinary or multidisciplinary theory. There has, however, been a rather common yet very peculiar relative absence of theory in contexts in which multidisciplinarity is celebrated as necessarily being 'enlightened'.

With the proclamation of the DMA in 2003, a still continuously growing DRAaM research industry emerged. With the institutionalisation of postgraduate teaching in this field, academic research grew in the form of master's and doctoral theses as well as staff and postgraduate student academic papers. Again, the matter of concern here is the trend by which the production of multidisciplinary research overtly framed as DRAaM has increased significantly and not the fact that such multidisciplinary work exists. One might easily make the argument that traditional disciplinary boundaries, especially within the humanities, have become untenable, as cross-pollination between disciplines is virtually inevitable in any substantive scholarship. The problem with much of the so-called multidisciplinary DRAaM research, however, boils down to a dearth of theory in general, be it disciplinary or multidisciplinary theory. Stated differently, it is not necessarily the multidisciplinary nature of the research which makes it theoretically weak. However, as it appears that this lack of significant conceptual analysis is far more common in such research contexts, we are confronted with a fundamental contradiction. Instead of adding to our understanding of the disaster problem by traversing the confines of disciplinary knowledge, the opposite is taking place.

The lens through which we view data has an impact on the quality of our analyses. This is why triangulation of theory is often included in research methods textbooks in reference to data validity (see Neumann, 2006:98). Without diverse analytical tools to draw on, a researcher will form a less precise interpretation of a particular matter. More conceptual tools could produce more precise prose based upon a nuanced and carefully considered analysis. Unfortunately, it seems as though a number of governing forces have coalesced at critical nodes within South African DRAaM knowledge production and in such a way that theory has often been side-lined. This is nowhere more evident than in the general consensus on the inherent value of multidisciplinary work and its implied, when not formally articulated, premise that there are inherent limits to disciplinary work (cf. Stanganelli, 2008:109). For our immediate purposes, we once again need not look further than the NDMF, specifically the definition of DRM cited in South Africa (2005:2), as 'integrated, multisectoral and multidisciplinary'. This approximate point of view has a lot of merit. Nevertheless, this does not mean that, given an appropriate context and focus, disciplinary disaster research cannot be meaningful. It also does not mean theory should be relegated partially or, as has frequently happened, manoeuvred almost completely out of view. Prevailing discourse/slogans/rhetoric (to reiterate, not entirely without merit) such as 'disaster management is everybody's business', have no doubt had an impact on how DRAaM academic programmes and the specialist units within universities were and continue to be viewed. This includes how the academics associated with these programmes choose to define themselves. Most likely, however, there are numerous complexly related causal mechanisms at play and, as such, these processes would have manifested as both dependent and independent variables at some point. One might even go so far as to argue that the very notion of causality, for example in light of Chapter 3's analysis, is highly problematic. Though that proposition might stimulate meaningful reflection, we cannot follow the defeatist course already encrypted into such a belief. Instead of acknowledging this complexity and the likely realisation that local context matters, researchers have often moved in an opposite direction, side-lining theory building and often relying exclusively on the handful of rudimentary concepts that is the DRAaM code and operationalised through the technocracy encrypted within the NDMF.

Multidisciplinary research may quite feasibly yield many useful results, in no small measure due to the obvious fact that such research potentially draws on insights from various disciplines. In some cases, unfortunately, DRAaM has served as a terrain between the various disciplines, where instead of drawing on each discipline's strengths, the research rather suffers a fatal lack of identity and purpose, as none of the disciplines contribute to any significant extent, let alone work together in yielding insights superior to traditional disciplinary research. Here 'multidisciplinarity' becomes an end in itself or a means to other agendas contingent upon the vagaries of professional patronage in our contemporary epoch of quantifiable output-based academia. Disaster risk assessment and management's apparent lack of theory, then, conveniently harbours the excuse that there is no/little theory to draw on by researchers who are unwilling, or who have insufficient time, to read more broadly. This may be due to immense teaching, research, consultancy, graduate supervision and the like responsibilities. Sadly, in such a case, the very objective of working across traditional boundaries is defeated, not to mention the steady erosion of this individual's essential academic ability. The implicit argument then is that there is no sense to be made of this vast terrain, just a brief moment ago fervently associated with numerous, established disciplines, hence the argument for multidisciplinarity. Let us now finally consider an example illustrative of the discussion thus far.

The work of Tempelhoff et al. (2009) draws together the labours of researchers from history, political studies, public administration and tourism management. This journal article, based on a research report, investigates the impact of flooding in the Garden Route from 2004 and 2005 on local residents, with specific focus on the tourism industry. The paper includes an historical discussion of flooding, focusing on incidence in the area, but no theoretical framework or literature review on similar contexts. The result is a paper that is very descriptive but lacking in analysis. Furthermore, instead of developing a common understanding, drawing together contributions from the researchers' various disciplines, each descriptive disciplinary discussion rather forms a silo within the paper, though the word 'disciplinary' is an unsatisfactory descriptor of such analyses, largely devoid of discipline-specific conceptualisation. Unsurprisingly, the conclusions and recommendations are more a collection of statements than the outcome of summative/reiterative arguments. This is because the article generally displays such a characteristic. Arguments are not developed and linked to a central analytical thrust. Each discipline seems to contribute some description on a theme, typical of that discipline, though barely (if at all) linked to established conceptual tools. In this instance there can be no meaningful practical recommendations, because the sociological and ontological 'relevance' or significance was never engaged with through conceptual discernment. For its descriptive superficiality, the article cannot arrive at a coherent definition of the research context in relation to the pertinent problematic of flooding. By, for the most part, merely describing, the authors do not acknowledge that this is a problematic in the first place.

Essentially, this study and others like it remain incomplete. It offers no coherent conceptual argument that is sophisticated yet focused enough to yield meaningful recommendations or that at least explains the practical implications of the analysis. Discussions are broad and disparate, an obvious result of the multiple disciplines not working together but alongside one another, executing multiple and superficial research projects on a more or less similar though very general topic. As a result, the study also lacks data validity. There can be no measure for data validity in such *atheoretical* analyses, as the very term 'validity' only finds meaning in relation to concepts established or newly generated. The analysis lacks the sociological significance which would have been offered through various potential conceptual markers, directing analysis and thus by definition producing knowledge. This hypothetical knowledge would be contextualised and based on an appropriate symbiosis between concept and data. Only once these criteria have been met, when there is sufficiently nuanced understanding of the dynamic processes related to salient concerns, can we speak of 'relevance'. We should at least add the words 'practice', 'praxes', 'potential interventions' and so on, as all of these necessarily imply meaningful understanding as a prerequisite. It appears that much of what is explained immediately preceding is linked to the intertwined nature of academia and practice and the 'irrefutable' relationship between time and money.

Research, regardless of the terms under which it occurs, and the quality of the results produce data to be exploited along further income generating courses. As far as academic research is concerned, the South African state subsidises outputs far more than it does inputs. 'Accredited' peer-reviewed journal articles are by far the most significant subsidy-generating output. Researchers, in this regard, often turn research reports into journal articles in order to maximise output, which is obviously conceived of quantitatively. For various reasons – many of which might be found in the constraints discussed in this chapter and the one before – authors often make little effort to add to the scholarship of the particular commissioned report in the process of converting it into an academic article.

This explanation, though certainly valid, remains unsatisfactory. Even commissioned research reports, as many successfully do, ought to conceive of and engage with the problem in a meaningful manner. If this is not possible, it is a bad idea to take on the project. Providing 'quick and dirty' answers diminishes the reputation of academia, not to mention the academic concerned, and does not serve the objective of the research, in other words to produce additional insight. Here one could simply refer to the discussion on validity and sociological significance as a requisite for knowledge production and the perils or even danger of praxes based on poor knowledge. It seems, however, that in the act of knowledge production, what has been the core characteristic of the academy as a societal institution has also been instrumentalised. Where state departments and non-governmental organisations (NGOs) commission questionable projects in order to simply spend a budget or where the academic publishes an article to build a career, the fruits of professional labour have become a means to different ends, more or less removed from that of producing new knowledge. As with the previously cited example of additional qualifications as a requisite for promotion in the workplace, the point of research has become less about generating meaningful insight or the competence a publication is meant to demonstrate. To some extent, it seems, any research report or article will do, as long as it generates the desired income.

Another common manifestation of this lack of meaningful conceptual analysis in DRAaM research is where legislation and theory are confused and, as a result, where the former is used as if it were the latter. By not drawing on substantial theory, and sometimes none at all, these analyses regress into policy fetishism, typical of the pervasive self-referential cycle characterising much of the relationship between DRAaM academia and practice.

Legislation as theory

Here again we find reason in the service of legislation telling of public administration's influence on DRAaM's South African institutionalisation. The basic formula followed in such research is to focus on legislation in the literature review, typically of a deductive analysis. In this review of statutory frameworks, authors would usually spell out a set of legislative requirements in managing a particular issue. In terms of DRAaM, the NDMF remains the most commonly used benchmark for analyses, essentially measuring legislative compliance.

Examples of this type of study include investigations into the placement of disaster management centres at various levels of government, whether municipalities in the Eastern Cape meet various NDMF requirements in a given area, institutional capacity for DRAaM (in accordance with KPA 1 of the NDMF), and the existence of DRMPs and compliance (seemingly in general) with the DMA (cf. Ddungu, 2008; Dlamini, 2011; Van Riet and Diedericks, 2010). In some instances the authors also draw on UN policy documents or field guides, likewise hardly the underpinnings of substantive scholarly analysis. It is not necessarily a problem focusing an applied research project on activities found in a field guide or policy document. However, when the practical tools and/or legislation are not situated within a broader conceptual framework, the very notion of knowledge production is fundamentally compromised. What the authors of these studies do not fully appreciate is that policy can never be fully universal. It can only follow from an adequate contextualised ontology punctuated by careful analysis of the key dynamics pertaining to the research objectives that conceptual analysis (and of course field research) delivers. It cannot be the other way around. To invert this relationship between scholarship and policy is to discard history. Without history, there can be little contextualised analysis, and all that remains is the supposedly universal code, an imaginary panacea unhindered by time or geography.

Legislation was never intended to be an analytical tool. It rests on a set of assumptions, which should be spelled out and reflected upon when used in scholarly analyses. Otherwise the analyses not only lack the fundamental self-awareness becoming of scholarship, but they fail to meaningfully engage with the ultimate reason for the legislation being enacted in the first place. As a result, these analyses are rendered decidedly ahistorical and conservative, as no other findings than those in some way in favour of current policy are possible. The legislation is essentially treated as if it was permanent and above critique, and the quality of the academic knowledge produced is questionable, as researchers virtually exclusively use conceptual tools extracted from the previously mentioned self-referential system of officials, consultants and researchers. Such conclusions of course logically fuel a more general conclusion previously encountered and dismissed, namely that we have excellent legislation but that we have equally poor implementation. This conclusion is unjustifiable in the absence of a justification of the legislation's appropriateness to a context which, to reiterate, can only be done by suitably conceptualising that context. There might, for example, be very good reasons municipalities do not comply with legislation, although insights which potentially provide far more precise understandings than merely asking whether or not a municipality complies with legislation remain untapped.

As Chapters 2 and 3 revealed, there are numerous discourses and bodies of literature pertaining to disasters without being framed as such. These include literature on housing, access to electricity, unemployment and social justice in general. Nonetheless, DRAaM has been and continues to be structured in such a manner that very little insight and few analytical tools – as a participant pointed out during an interview – are drawn from these older and far more developed bodies of literature in order to understand the problem of disaster. Being 'at risk' is not the only affliction many South Africans have to contend with. It might not even be the most important affliction. Furthermore, particular forms of danger might be something many are willing to tolerate in order, for example, to live close to work, where the cost of daily transportation over significant distances is unaffordable. How does knowledge of such social and political-economic dynamics affect citizens' daily decision making? These sorts of questions remain largely unanswered, if not stubbornly ignored, by researchers fixated with management from a distance in devotion to questionable interpretations of a policy document.

Consultancy and knowledge production

An analysis of consultants' reports reveals the various assumptions and dispositions both produced by consultancy and constitutive of the governed intellectual space that is DRAaM. In South Africa DRAaM is primarily governed by the first discourse discussed that follows namely, expert knowledge claims. Broad acceptance of such claims to authority is the necessary condition on which privatised DRAaM rests. Other expert and industry-constituting discourses are, to continue the genealogy of Chapter 1, in the service of unauthentic ends. Experts deploy strategic words and phrases in order to further justify their standing with clients and colleagues.

The most imperative DRAaM actor-constituting imperative seems to be the maintenance of the professional, the expert who leverages the industry code in diverse contexts. To draw on Foucault, *what power does* is to subject any and all additional actor-constituting actions to instrumental-ity in the service of a professional identity. The South African DRAaM industry hinges upon the recognition of consultants' specialised knowledge claims. This is a characteristic of the general neoliberal government insti-tuted post-apartheid. As a general international policy turn, itself a result of prevalent neoliberal modes of governmentality, NPM offers a haven from which various private-sector service providers can launch their pro-fessional pursuits.

These specialised knowledge claims are based on privileged access to the commodified code. Even though the discussion in Chapter 3 merely scrapes the surfaces of many of South African society's troubling attributes, it does reveal the diverse interacting and often overlapping layers of suffer-ing that many have to endure. The sheer complexity of the matter implies that the afflictions of structural violence cannot possibly be explained by consultants with strict time frames and budgets. Consultants, however, do not need intimate knowledge of a research context to project authority based on specialised knowledge. They market the code and the supposed promise it holds for unlocking particular market-mandated forms of understanding, by tapping into the explicit and tacit knowledge of key informants, local residents or, as is too often the case, reinterpretations of standardised data-sets (e.g. census data and GIS databases). The following analysis explains how this application of professional code cannot make up for the hitherto lacking-in-depth context-specific knowledge.

Legislative compliance

Consultants' livelihoods depend on helping clients meet their legislative mandate. Adequate remuneration also depends on a beneficial relationship between time and income. As a result, legislative compliance has mani-fested as a malleable actor and industry-constituting discourse. Interpre-tations of 'compliance' in extremely limited and technical terms produce superficial action, largely reduced to an essentially bureaucratic exercise, largely focused on producing generic reports and institutional structures. Many consultants' reports appear to be based on the structure of the NDMF, often to the extent that vast proportions of the text are identical to sections of the NDMF. In other cases reports follow the same headings found in the NDMF but offer a distilled version of that document. Significantly, often not much is achieved in addition to producing these reports.

Legislative compliance pertaining to DRAs is typically interpreted even more narrowly, only including some of NDMF's propositions and require-ments. For instance, in most cases consultants ignore 'ground truthing' or gathering data, as it is worded in the NDMF, at the 'community' level (South

Africa, 2005:37). In some cases consultative meetings with municipal officials are even stated as drawing on 'indigenous knowledge'. The authors
thereby leverage the authority often associated with this problematic term
without in any way dealing with the populace. This use of the term 'indigenous knowledge' might be brought on by ignorance or even a blatant dis-
regard for more or less established denotations. Regarding such definitions,
Laurie et al. (2005:14) note,

> In the emerging academic literature on indigenous knowledge, most
> definitions are linked to natural resource management, knowledge
> about a specific territory and/or knowledge held by a particular group
> who are assumed to live in a bounded geographical space.

Bureaucrats as key informants cannot possibly have such knowledge about
an entire municipality, especially since they most probably live in a com-
parably safe, middle-class suburb. Consultants do not need to understand
all of the words they use and the potentially adverse consequences of such
usage. They merely have to *perform* the act of DRAaM by underlining key
markers forming the general scaffold of the field and associated notions
of 'best practice'. Moreover, in pursuit of measurable performance targets,
the relevant municipal department has demonstrated compliance by hiring
someone to produce evidence in this regard.

In other cases, the degree to which even limited interpretations of legislative compliance can be met is limited to the skills set found within the
consultancy, for example business continuity management, GIS, or some of
the social sciences. Therefore legislative compliance, and in particular 'multidisciplinarity', seem to mean different things in different contexts. This is
perhaps understandable, as the industry is made up of private companies
or third-money-stream–generating units based at universities. Appointing
a more diverse skills-set or subcontracting such skills will almost certainly
threaten profit margins. This has an obvious implication, in particular for
notions of 'multidisciplinarity' often ascribed to DRAaM and, as such,
another constitutive discourse.

Multidisciplinarity

The NDMF and numerous publications emerging in South Africa from
2003 and following global precedent state that 'DRM is an integrated multisectoral, multidisciplinary' activity (cf. Roth and Becker, 2011:447; South
Africa, 2005:2). As this is constantly reiterated, one might expect practitioners to take multidisciplinarity seriously. This is not the case, as multidisciplinarity is relatively insignificant as a governing discourse and as an

influence on industry practices. Rather and much similar to the case of legislative compliance, multidisciplinarity as a discourse is subordinate to power exercised through the more central features of capitalist governmentality.

We must bear in mind this discussion is partially attributing superficial compliance to consultants' need for financial survival. The result is an overwhelming focus on management and the convenient equation of DRR with administrative-legal requirements, used by bureaucrats to demonstrate 'performance'. Nevertheless, the rhetoric that DRM is a multidisciplinary activity remains. The following quotation is illustrative of the point:

> [B]ecause any risk assessment, [it] does not matter what you are going to do, you must get input from engineers. You must get an environmental scientist. You must, if it is weather-related, get weather data and have someone analyse it. You must have a sociologist. And then you must have the ability to add it together and calculate it analytically, without thumb sucking. So you have to be able to work in a team.
>
> (Author's translation)

Similar to many others in the field, the participant quoted here appears to have much ambition for DRAaM. Drawing together so many diverse skills and then synthesising them into coherent sets of actions would have to be an inordinately challenging undertaking. Multidisciplinary teams largely comprised of employed academics have had much difficulty in DRAaM research in South Africa, as is evidenced in the foregoing. Academics, even those with one foot in industry, arguably have more time to allocate to research than consultants. At least in theory, cross-pollination between teaching, research and consultancy implies a somewhat more favourable funding structure. Yet some multidisciplinary academic research teams have produced rather incoherent work. It therefore seems unrealistic to expect that consultancy firms possess or will contract such a diverse set of skills when their primary concern is the 'hard interests' of their firm's continued financial viability (see De Waal, 1997). It should therefore be no surprise that this multidisciplinarity is not reflected in consultants' reports. Even where reports have multiple authors, it is very rare that authorship extends beyond two or three disciplines. Some consultancies have only one person working on DRAaM. Hence, very often reports supposedly produced by externally contracted specialists are in fact produced by generalists.

Technical expertise based upon the scientific method does have an obviously indispensable role to play in knowledge production related to disaster. For example, studying sinkholes without consulting a geologist or physical infrastructure without consulting a civil engineer or working with people at a very low level of analysis and without some concept of social science research practice would be irresponsible. Unfortunately this is exactly what is happening, which raises the question: if DRM is a multidisciplinary activity, then why is this often not reflected in consultant reports, and why is a product not based on all of the relevant scientific principles still marketed as such?

Science

The South African DRAaM industry members seem to fixate on procedures and technology as criteria for data validation and as markers of expertise. Industry is largely ignorant of the conventional requisites, if not defining characteristics of scientific knowledge, such as the trustworthiness or validity of data. To paraphrase from any research methods textbook, for data to be valid, measures should in fact measure what they are intended to measure. Yet many reports draw on aggregate datasets, such as census data, which in some instances were still being used as a 10-year-old dataset, and land-cover or land-use maps, from which questionable extrapolations are made. These practices surely pose significant challenges to even some of the most liberal conceptions of validity. Understanding the emphasis on technique above substance as a means of projecting professionalism should shed more light on matters of validity as an essential, defining characteristic of science. Compare the following examples.

The equation below is taken from a risk assessment report. It was used to measure the impact of a particular hazard on an area (Eden District Municipality, 2006).

$$I = SI \times [SoV + EcV + EnV + StV]$$

What each letter in this equation is meant to refer to is frankly not all that important. What is important is the fact that such technical parameters were used to distil complex matters, the likes of which occupy much of Chapters 2 and 3, to a single score.

One consultant, in reference to the prevalence of DRAaM outsourcing, noted, 'It is not that difficult. You can actually do a risk assessment on an Excel sheet' (Author's translation). This participant believes that local government officials can do the risk assessment themselves by using Microsoft Excel. Therefore, the problem of reducing complex reality to an excessively simplified technical exercise endures. Moreover, with both examples, the parameters consultants use in order to fill in the missing values are based on subjective and often arbitrary measures, which are then quantified. Municipal officials might for example be asked to allocate a value between one and three for the severity or likelihood of an event. The consultants would subsequently calculate the values, reducing them to a consolidated score of, for instance likelihood, severity or even risk. To reiterate, these are subjective measures, and the values for each are produced by consultants who typically do not know the area being studied at all, based on municipal officials' ratings of a very limited number of distilled and preconceived variables. Therefore, while the seemingly sophisticated quantitative methodology might look impressive to the client, it merely obscures the subjective nature of inputs. I do not mean to make a case for 'objectivity' in social science, which is arguably a false notion. However, openly subjective analyses will at least present substantiated arguments that opponents could consider and respond to. Subjectivity in such cases is acknowledged, not buried under questionable formulae. Fixation with quantification extends to the most elementary of methodological aids, where it is no longer method that is the object of fixation but technology, and this fixation with technology is nowhere more evident than with the status ascribed to GIS.

The NDMF requires hazard mapping as a means toward understanding disaster risk (South Africa, 2005:34, 48, 63). This requirement is often interpreted in such a manner that DRAs become GIS driven, a practice we may contrast with instances in which GIS is simply used to add insight into pertinent disaster- and danger-related issues (Van Riet, 2009). These desktop risk assessments often use land-cover and land-use maps to deduce the potential for particular adverse events. In other words, particular types of land cover might be considered more flammable than others and assigned a score, which then informs the prioritisation of areas threatened by veld fires. The problem with such analyses is that they lead to simplistic causal explanations, ignoring social relations, in addition to other biophysical phenomena, not *operationalised* based on the data at hand. They also ignore the fact that land use and land cover change continuously. As a result, reports become dated relatively quickly. However, this does not seem to matter, as the mere presence of maps often becomes a performance indicator in itself. Often most of the assessment is presented in the form of maps, with little and limited explanatory text. Sometimes consultants draw parameters around a given hazard, such as a rubbish dump or an airplane flight path. In these cases proximity is equated to vulnerability with virtually no additional discussion, including explanations and justification of the distance chosen for the parameters. Beside the fact that they are often based on outdated data, the analyses typically exclude relevant social, political and economic dynamics.

Consultants actively market the visual appeal of maps associated with GIS, probably because it is mentioned in the relevant policy documents but even more importantly because such procedures are far more cost effective than actual field research. Analyses such as these allow for quick movement between projects and, as such, optimal income. Nevertheless, the reference to hazard mapping in the NDMF provides a convenient avenue to commodification, perhaps based on an appeal to the senses but also impressing with supposedly sophisticated technologies, which validate the 'expertness' of consultants.

Subsequent to an initial discussion of hazards in a particular setting, risk assessment reports typically elaborate on vulnerability, though often as 'vulnerabilities'. Sometimes there are discussions of capacity, which in a similar fashion are framed as 'capacities'. As has been argued elsewhere (Van Riet, 2009), be it in a more positivist vein, discussing hazards and vulnerability in silos would not likely facilitate speaking to the actual problem. Furthermore, as the word 'vulnerabilities' as opposed to 'vulnerability' suggests, these aspects are often treated as separate from one another. In other words, what is essentially a convoluted maze of complex and dynamic interactions is treated as a list of separate factors, in some cases even listed with bullets. This enables the de-politicising framing of dangers as matters for risk management. These factors can apparently be listed, and then all that is left is a few simple interventions drafted in a few hours for all risk management in the entire district.

Disaster risk assessments often begin from a problematic position, brought on by reliance on preconceived operational definitions of key concepts. These are either in the form of a master list of hazards that participants select from or arbitrary categories such as 'often', 'occasionally' and 'seldom' or 'high', 'low' and 'tolerable', pertaining to the levels of probability or severity of hazard impact, each assigned a score. Reports typically offer no explanation or even indication of how these categories are grounded in any sort of theoretically and empirically defensible logic, thus making the validity of the resultant findings highly questionable.

With the master list of hazards used for risk assessment, potential dangers in a given area might be excluded from the very beginning. The preconceived list is essentially imposed from outside, based on presumption and failure to acknowledge that 'we do not know what we do not know' (Freudenburg, 1996:48–49). Only this is presumption articulated from positions of power based on recognised expertise and as such fashioned into knowledge. In light of the complexity and need for sensitivity to local contexts, which has been a central argument especially in the last two chapters, a more open-ended approach seems far more likely to yield results that are relevant, in both a sociological and a practical sense, than any prior narrative imposed on society from above. The alternative, following the participatory turn in development studies, took some time to arrive locally. When it did, participation, too, turned it into an unsatisfactory performance of professionalism.

Community-based disaster risk management

The NDMF (South Africa, 2005:18–19) conceives of DRR as a community-driven process. The document states that communities should be involved in developing risk profiles and that risk reduction project teams must include community members. This approach to DRAaM is, however, largely ignored by industry at large, judging by consultants' reports, especially prior to 2009. Nevertheless, CBDRM is one area where the dynamics of structuring power seem to be shifting, although only partially and very slowly. Generally, the consultants interviewed noted CBDRM as good practice. This is perhaps a first finding with regard to CBDRM; despite rhetoric to the contrary, DRAaM at very low levels of analysis, though more noticeable in industry discourse of recent years, is still not a very common practice. Community-based risk assessments and subsequent management practices are necessarily more expensive, as they are more time consuming and include more logistical complexity. Furthermore, given time and financial constraints, instrumentalisation is the logical consequence, whereby the performance of CBDRM through an essential suite of participatory techniques becomes an end in itself. The client may then proclaim doing CBDRM and report this to a manager, regardless of the substance afforded to the concept.

One consultancy executes a form of risk assessment, which the employees define as community based, as a standard matter of practice. The work entails a threefold process. First, they consult municipal ward councillors. According to the senior DRAaM professional for the consultancy in question, the councillors are meant to provide initial information regarding the types of hazards found in the ward. This is followed by discussions with traditional leaders, chosen as informants for similar reasons and consulted with similar objectives in mind. Finally, 'random members of the community' are consulted on an ad hoc basis, though older members are still preferred. The focus throughout is on what has happened in the past in the ward, as opposed to what has not necessarily happened but may happen. The risk assessment document produced includes various ward-level discussions of no more than one page each, including a bulleted list of hazards found in the area. The 'analysis' provides a general overview of the ward, including livelihoods and demographics.

This approach is questionable on at least two grounds. First, the knowledge generated is exceptionally thin and standardised. The text generated lists known threats but adds nothing about why and how these affect the area. It is difficult to conceive of disaster-mitigating measures being developed based on a list and a single paragraph describing rather unrelated demographic characteristics. Many would in fact argue that, for its lack of vulnerability analysis, it is by definition not DRR. Second, the use of the terms 'community' and 'community-based' may be questioned, even by those who accept the terms as relatively unproblematic. Municipal wards are more or less delineated based on population size. These wards also tend to be quite large geographical units, possibly too large a level of abstraction for CBDRA. Most likely, many wards will include numerous groupings of households with different dangers relevant to each, especially but not only when one considers social, political and economic dynamics, something this particular consultancy does not consider. This particular approach to DRA, therefore, arguably defeats the primary objective of conducting work at community level. Finally, calling this form of information extraction 'community based' seems misleading. These projects, subject to limitations of the 'time equals money' rationality, are brief interventions which tend to yield very limited insight.

Van Riet and Van Niekerk (2012) reported on a large-scale DRAaM project in the Dr Kenneth Kaunda District Municipality in the North-West Province and found that the strict time frames, which are the very nature of consultancy work, led to significantly compromised CBDRA data. The project's objective in addition to DRA was to build capacity for DRA at municipal level in order to help end the cycle of municipal dependence on consultants. Municipal officials were actively involved in all facets of the project. Unfortunately, strict time constraints meant that 18 different sites were assessed within three months, creating various logistical complications and bottlenecks, most acutely experienced with regard to writing up analyses. As a result, the data tended to be thin, and reports tended to be fairly standardised.

Participatory research has value, first as it potentially provides better data, collected at a low level of analysis. Second, it has value because it provides an opportunity for meaningful participatory decision making and strong good-faith relationship building between state and society. If none of this materialises, then the practice should be questioned. Van Riet and Van Niekerk (2012) found that this would unlikely materialise when projects were outsourced. By outsourcing what should be a tool for participatory democracy, the complex political issue pertaining to relative access to power and relative affliction by the adversities of modernity, which belongs in the realm of the normal politics of deliberation, is de-politicised. It subjects the matter of concern and ordinary citizens, often with legitimate concerns, to an array of standardised interventions in the name of risk assessment and management and the hard interests of practitioners. Re-politicisation occurs through the acts of re-configuring power dynamics between the expert facilitator and the participants who are assessed. The facilitator leads participants through the various participatory activities designed to extract data amenable to DRAaM code, essentially a unilateral framing of citizens' concerns, exclusively based on a particular imposed professional agenda. All alternatives have been framed out of the picture from the outset. While this practice might be excused when tangible results ensue that benefit participants, this type of objectification is less forgivable when the result is little more than a paycheck for the facilitator and possibly a performance bonus for a bureaucrat. However, the relative impotence of CBDRA is not due to any overt sinister plot on the part of facilitators and their client, local government, but rather due to how DRAaM confines and directs discursive flows as a component and a product of dominant modes of structuring throughout contemporary society. This celebration of professionalism founded upon the aforementioned constitutive discourses has and continues to constrain the potential for emancipatory praxes.

As with other forms of risk assessment, the fundamental modernist intellectual *cul-de-sac* prevails, following from the presumption of DRAaM's universality. This presumption forms the basis for standardisation in general and the (up to now) inevitable instrumentality of these standardised performances. In the case of CBDRA, as with other forms of DRA, it is still not the expert knowledge of the study area that is sold or sought after. Rather, it remains the promise that applying the universal expert code will unlock enough understanding to achieve a particular version of legislative compliance.

Conclusion

The emerging industry of theses, dissertations, papers and articles framed as DRAaM attests to the same legislation-reifying instrumentality described in Chapter 4. Much of the DRAaM scholarship fails in its fundamental mandate: to provide meaningful interpretations of the disaster problem. Essentially conservative scholarship, therefore, has become a mere manifestation of the disjuncture between the ideals of disaster reduction, conceived in the 1970s, and its institutionalisation through policy convention globally. In contrast to the conclusions from especially Chapter 3, the DRAaM literature reviewed, though paying lip service to the link between notions of poverty and vulnerability, still seems to leave unquestioned this poverty or the structural violence of the everyday which shapes vulnerability. Many would argue that poverty and vulnerability are not synonyms, while arguing that it is normally the poor who suffer most from disasters and are more frequently affected by disaster. There does, however, lurk a clear oversight in this reasoning. Structural violence as a deliberately overt political statement and conceptualisation, similar to marginalisation though more precisely focused, still subsumes states and processes described by many DRAaM authors as vulnerability. But because structural violence or even more exclusively materially oriented notions such as poverty remain largely unquestioned, references to structural change of the largely illegitimate political economy are nowhere to be found. Such an omission is understandable in the realm of consultancy. Within the context of academic knowledge production, however, such an omission is unacceptable. Academics, in defining the broader problem, are surely 'in the business' of offering explanations that extend beyond immediate practical utility and micro-level interventions. But please note the language being used: *immediate* practical utility. In order to achieve meaningful change, it must once again be reiterated that one needs to provide first and foremost a useful understanding of the situation in need of change. Understanding of this nature, as it has been argued, requires systemic change *and* truly contextualised micro-level insight. The need for structural change is not a trade-off with micro-level action. Escobar's (1988:439) work on popular grassroots movements suggests that the latter may often be a means to the former.

Similarly, regarding consultancy, one may ask: is it really important that municipalities comply with the NDMF, or should we rather focus on understanding the problem of disaster and its relevance in a particular context? Institutionalised modes of codified knowledge production are highly problematic, as they are necessarily closed off to all possible interpretations of a given context, bar one. Even in cases where disaster is a priority concern, reliance on consultants to conduct much of DRAaM makes it difficult to address the problem. These projects often translate into knowledge that is unusable and state–citizenry dialogue that is limited at best, making it hard to justify the amount of money spent.

Expert interventions are typically informed by cultural interpretations of risk, which I have consistently challenged in the last four chapters. In other words, danger regardless of practitioner intent, by default, is taken as emerging from ignorant locals who do not know any better. Carrying this type of analysis to its logical conclusion, what is essentially required is more citizen awareness. Recommendations from consultants' reports, in addition to the technical formal legislative requirements (convenient interpretations) discussed earlier, and in line with Enabler 1 of the NDMF (information management and communication), do indeed tend to propose public education for disaster risk reduction (cf. Alfred Nzo District Municipality, 2011:56, 59, 61, 62, 65, Pixley ka Sema District Municipality, 2010:9–10). I do not mean that the likes of awareness campaigns may not fulfil *a* function. That would be an irresponsible and untenable position. Macro-level social change is unlikely to be a short-term undertaking, nor would it affect all at the same tempo, nor will it even affect all in a positive manner. Acknowledging the potential utility of such cultural interpretations of risk, unfortunately, comes at potentially too great a cost, where policy makers give too much credence to this type of intervention, thus elevating generic and partial interventions to the status of panacea.

Note

1 An ideal type is a hypothetical concept abstracted for the purpose of analysis. These categories do not correspond perfectly with aspects of reality. They are, however, useful in facilitating analyses. In this case the distinction between academia and practice is not perfect. Hence, they are conceived of as ideal types.

References

Alfred Nzo District Municipality. 2011. Municipal Disaster Management Plan. Version 1. Online: http://www.andm.gov.za/Documents_Download/Plans/Documents/ANDM%20Disaster%20Management%20Plan%20Version%201.pdf (Accessed 21 June 2011).

Coetzee, C. 2010. *The Development, Implementation and Transformation of the Disaster Management Cycle*. Master's dissertation. North-West University, Potchefstroom.

Ddungu, P. E. M. 2008. *An Evaluation of the Disaster Management Function of Municipalities in the Gauteng Province*. Master's thesis. University of the Witwatersrand, Johannesburg.

De Waal, A. 1997. *Famine Crimes: Politics and the Disaster Relief Industry in Africa*. Oxford/Bloomington: James Currey/Indiana University Press.

Dlamini, P. 2011. *Evaluating the Implementation of the Hyogo Framework for Action in the Kabokweni Location: Views from the Frontline Perspective*. Master's dissertation. North-West University, Potchefstroom.

Eden District Municipality (EDM). 2006. Eden District Municipality: Developing a Disaster Management Plan and Conducting Risk Assessment for the Eden District Municipality. Unpublished Report EDM. George.

Escobar, A. 1988. Power and Visibility: Development and the Invention and Management of the Third World. *Cultural Anthropology*, 3(4): 428–443.

Freudenburg, W. R. 1996. Risky Thinking: Irrational Fears about Risk and Society. *Annals of the American Academy of Political and Social Science*, 545(1): 44–53.

Kasie, T. 2009. *Vulnerability to Food Insecurity in Three Agro-Ecological Zones in the Sayint District, Ethiopia*. Master's thesis. University of Cape Town, Cape Town.

Laurie, N., Adolina, R. and Radcliffe, S. 2005. Ethno-development: Social Movements, Creating Experts and Professionalising Indigenous Knowledge in Ecuador. *Antipode: A Radical Journal of Geography*, 37(3): 470–496.

Mavengere, C. A. 2011. *Human Agency and Everyday Risk: Comparing Household Protective Measures for Children in Ward 7, Epworth (Harare), and Samora Machel Philippi (Cape Town)*. PhD dissertation. University of Cape Town.

Mgquba, S. K. and Vogel, C. 2004. Living with Environmental Risks and Change in Alexandra Township. *South African Geographical Journal*, 86(1): 30–38.

Neumann, W. L. 2006. *Social Research Methods: Qualitative and Quantitative Approaches*. Hoboken, NJ: Pearson.

Pixley ka Seme District Municipality. 2010. Pixleyka Seme District Municipality: Disaster Management and Contingency Plan. October 2010. Online: http://www.pixley.co.za/documents/Policies/Disaster%20Management%20Plan%20 10032010.pdf (Accessed 31 May 2013).

Roth, A. S. and Becker, P. 2011. Challenges to Disaster Risk Reduction: A Study of Stakeholders' Perspectives in Imizamo Yethu, South Africa. *Journal of Disaster Risk Studies*, 3(2): 443–452.

South Africa. 2005. *National Disaster Management Framework*. Pretoria: Government Printer.

Stanganelli, M. 2008. A New Pattern of Risk Management: The Hyogo Framework for Action and Italian Practice. *Socio-Economic Planning Sciences*, 42(2): 92–111.

Tempelhoff, J. W. N., Van Niekerk, D., Van Eden, E., Gouws, I., Botha, K. and Wurige, R. 2009. The December 2004-January 2005 Floods in the Garden Route Region of the Southern Cape, South Africa. *Jàmbá: Journal of Disaster Risk Studies*, 2(2): 93–112.

Van Riet, G. 2009. Disaster Risk Assessment in South Africa: Some Current Challenges. *South African Review of Sociology*, 40(2): 194–208.

Van Riet, G. and Diedericks, M. 2010. The Placement of Disaster Management Centres in District, Metropolitan Municipality and Provincial Government Structures. *Administratio Publica*, 40(2): 155–173.

Van Riet, G. and Van Niekerk, D. 2012. Capacity Development for Participatory Disaster Risk Assessment. *Environmental Hazards*, 11(3): 213–225.

6 Multiple dystopias

Introduction

Notwithstanding the analyses of the previous two chapters, those working in DRAaM are not 'bad people'. Instead, as with any field of practice, DRAaM practitioners are governed by a set of enabling and constricting discourses. These have shaped a particular type of practitioner and practice based on consultancy, instrumental reason, much superficiality and questionable knowledge production.

This chapter adds an element not often included in critiques of development by considering how it affects practitioners. Thereby, I hope to in the first place demonstrate many practitioners' sincerity, noble intensions and critical introspection that are evident in much of the interview data and second to link the experiences of practitioners with the material covered, in particular in Chapter 3. There are many more or less direct negative consequences of DRAaM's institutionalisation in South Africa. The status quo leaves much to be desired and, as such, the manner in which DRR, development in general and associated practices of state building are approached requires extensive reconsideration. Conclusions on the latter are offered in the final chapter. The current chapter engages with the effects of DRAaM on practitioners in the wake of the structural violence characterising the lives of many intended beneficiaries.

Neo-*compradors*

Disaster risk assessment and management in South Africa has manifested as a type of barbarism, analogous to that discussed by Adorno and Horkheimer. Similar to apartheid South Africa, neoliberalism-infused DRAaM as a form of institutionalised instrumentality has produced or proven impotent in the face of multiple dystopias across the South African social, political and economic landscape. All of the work undertaken in the name of the vulnerable has been appropriated by the superficially standardising inklings of practitioners' quest for self-preservation.

The actor-constituting discourses discussed above serve as the basis for such self-preserving actions, as actors engage in activities that are considered generally accepted practice, in the process serving their material needs or hard interests while providing the ontological security that accompanies professional identity.[1] Professional identities also tend to inspire their own self-serving rationalities. These rationalities and corollary interventions have affected little more than formal municipal administrative procedures, though even such procedures often fail to endure. The result has been that much of the initial emancipatory sentiment of disaster reduction has been lost in pursuit of standardised administrative performance indicators. The implications for both the stated beneficiary and expert are severe.

As much of the contemporary South African dystopias, for the typically intended beneficiaries of DRAaM, are the focus of Chapter 3, the following discussion primarily focuses on practitioner experiences of distress. The dystopias encapsulated in the modes of suffering discussed in Chapter 3 of course remain the more pressing and urgent matters of concern. The DRAaM expert does very little to alleviate these forms of suffering. We can now add to these contemporary South African dystopias, by way of a different though more exact focus on the DRAaM expert. To this end the analysis covers two themes: What is conceived of as neo-*compradorism* in reference to the DRAaM industry and the *anomie* many within the industry experience.

The burgeoning industry has created new jobs in academia, the state and private sector, broadly in the name of 'the vulnerable'. Whether the activities stemming from these positions are truly in the best interest of these officially intended beneficiaries is questionable. The impotence of DRAaM and its professionals, in no small way stemming from a lack of dialogue with the official beneficiaries, suggests it mostly is not. Therefore, this particular segment of the development set might be described as neo-*compradors*, drawing a salary mostly from tax revenue amid a predominant lack of substantive success in making a meaningful impact. Here the analysis draws on the Latin American Dependency School of the late 1950s to the 1970s. The *comprador* class discussed by such authors as Baran (1957) and Frank (1970) is defined accordingly as a class of locals in developing countries, peripheral or satellite states who benefit from the economic oppression inflicted through unequal trade with the economic centre or 'core states' upon 'peripheral' states. This class typically earned a living from raw material exports, in ignorance of the adverse implications of unequal terms of trade between the satellite and core or *metropole* states, associated with an open, resource-based economy. They are a class of Global South capitalists in the service of the North wittingly or unwittingly. On account of favourable access to Northern knowledge-powers, they are the presence of the North within the South, not the South itself. For the purposes of the central argumentative line of this book, the global dimensions of the South African citizenry's oppression must be set aside to focus on some of the country's internal political dynamics. These neo-*compradors* or compradors in the age of development and post-industrial capitalism are a category of capitalist wittingly or unwittingly in the service of an illegitimate political economy, and they operate under the guise of alleviating oppression while doing very little of the sort.

Yet the research has revealed that practitioners are at the behest of expert knowledge claims and what this frame of reference renders conceivable and feasible action. Complicities are found scattered on a continuum between malice and ignorance. I certainly did not encounter any positions toward the former extreme. For some, momentarily or sometimes enduringly, seeing through the fog of relative power and privilege, realising their impotence can be disconcerting.

Anomie

For these practitioners dystopia manifests as *anomie*. As such, the barbarism discussed by Adorno and Horkheimer culminates in a form of alienation on the part of the initial protagonists, who to their own realisation become antagonists, favoured by the macro- and micro-discourses informing daily life. The choice of Merton's conception of *anomie* as opposed to the general Marxist idea of alienation employed by Adorno and Horkheimer requires some justification.

Merton's (1938) particular operationalisation of *anomie* implies a disjuncture between a specific society's prevalent goals and the available legitimate means by which to achieve these goals. Deviant behaviour may result from society-governed goals such as affluence, purchasing of products, owning a house and so forth, while there are insufficient legitimate means for everyone to achieve these goals. In the case of the DRAaM practitioner, one might argue that the community of practitioners prizes goals such as helping others while at the same time earning a living and acquiring possessions they may desire. While Merton studied a society, the USA, at the national level, this book analyses an epistemic community conceived of as an industry as a different type of society. Of course, all societies are comprised of individuals with particular needs and aspirations. These are not unimportant, though as a social scientist one is more concerned with widely shared goals within a defined society, how these are constructed and how they govern the field. Some of the industry's goals might require a professional to produce research reports that meet the requirements of the field, including enough of the markers of marketable expertise. Here the pursuit of ontological security and as such dominance over nature is the pursuit of industry- and individual-specific goals. The goals practitioners' pursue constitute the Foucauldian assemblage subject, including the sense of self, including more or less shared elements, which might become an object of instrumental reason. What makes this sense of self different from that discussed by Marx is that it can be defined in terms of a set of industry- or context-specific goals, which are themselves subject to and shaped by the dynamic disciplinary discourses and processes of governmentality within that field. Therefore, there is nothing universal or preordained about the substance of any of these senses of self. It must, however, be conceded that the interpretation in this book still rests upon an assumed need for ontological security.

Anomie may for example manifest as knowledge workers learn that in order to secure a contract, they have to compromise in terms of the quality of work. Their pricing needs to be competitive, and the capitalist logic of 'time equals money' takes precedence as the time spent on the project decreases in order to move onto the next. Alex de Waal's (1997) distinction between the 'hard' and 'soft interests' of actors in the development industry is pertinent in this regard. In his discussion pertaining to famine relief in Africa, he argues that hard interests relating to the survival of the NGO (or, in the case of the current study, consultancy/consultant) are more rigidly pursued. In other words, hard interests relate to the ability to continue with your business and to make a living. Soft interests relate to the more official objectives of the business or at least to the reasons for the field emerging in the first place. When placed under strain, De Waal argues, hard interests are privileged over soft interests. The link with Merton's theory of anomie is clear. When faced with limited means to achieve goals, soft interests might be neglected, often leading to significant distress on the part of the practitioner, for example manifesting as a lack of pride in her work. In contrast to the Marxian conception of alienation employed by Adorno and Horkheimer, this can be seen as stemming from the governing function of the socially constructed expectation prevailing in the particular geographical, societal and industry-specific context and not by virtue of some or other inherent 'human nature'.

It should be stated that the usage of Merton's conception of *anomie* still allows room for the Marxian notion of alienation from the fruits of your labour, for example when workers are not proud of the fruits of their labour. However, 'alienation from yourself' is reframed in terms of a

specific set of socially constructed objectives. This adaptation and devia-
tion from the work of Adorno and Horkheimer means that Marx's broad
sentiment, one of distress on the part of the individual agent, is still cap-
tured but in a more theoretically and empirically defensible manner. To
reiterate, the usage of *Dialectic of Enlightenment* is qualified by the fact
that this book is premised on Foucauldian ontology. This has been justified largely with regard to the greater precision offered by such a basis
for analysis. It should furthermore be emphasised that the substitution of
alienation with the more open ended and therefore precise notion of anomie does not compromise the appropriateness to this particular context of
the general narrative of the *Dialectic of Enlightenment*. This narrative is
essentially the tale of an initial idea meant to liberate, which degenerates
into instrumental reason and the eventual barbaric consequences that follow for both practitioner and beneficiary.

In order to demonstrate that many of the practitioners interviewed experience *anomie*, the discussion that follows provides evidence that practitioners experience a sense of alienation from the fruits of their labour. Said
alienation is a result of a cessation between goals and legitimate means with
which to achieve these ends. In other words, actors knowingly deviate from
practices 'they can be proud of' due to a lack of legitimate means with which
to achieve such respectable objectives. The result often is a product that is
deficient and, as a result, actors experience more or less significant distress
at various levels of consciousness. Due to a lack of conceivable alternatives
these illegitimate actions and their consequences remain.

I will deal with two types of goals only, namely to earn a living and
also to serve humanity. However, when the former is threatened, the latter is often sacrificed. As with De Waal's distinction between hard (institutional survival) and soft (serving humanity) interests of international
non-governmental organisations (INGOs), South African DRAaM practitioners ensure institutional survival before contemplating soft interests. For
many it seems these soft interests have completely fallen by the wayside.
By far the most significant cause of neglected soft interests is the competitive capitalist nature of the industry. As time is equated to money, survival
depends on speedily moving from one project to the next. Pricing the product low enough is extremely important. The interviews revealed that this
latter strategy is near universal. Interestingly, however, virtually all of the
practitioners interviewed blamed others for the practice of driving prices
down, feeling forced to quote an unreasonably low price.

This particular knowledge-based industry's defining procedures seemingly do not allow practitioners the time and budget to do meaningful work.
Nor, as has been argued, does the governing DRAaM code allow the governed to fully consider the broader dynamics impacting on and stemming

from material danger. Some of the following responses are indicative of consultants' distress:

> Look, unfortunately your time is measured in money. So there is a budget and you only have X amount of time to spend on the thing. For me this is frustrating. You don't feel as if you ever achieve anything you are 100% satisfied with, because there isn't enough time and money for it.
> (Author's translation)

> Most of the time we try to consult people, but the problem here is, risk assessment budgets are too tight. You have to do what you do.
> (Author's translation)

> Sometimes it's like . . . 'Have you finished? Okay, good job. Next . . .' We just move onto the next thing. I come out of every project with this kind of frustration.

One participant, who uses DRAaM work to supplement her family's income, after herself explicitly questioning the legitimacy of DRAaM as a field of study and practice, responded as follows: 'There is no one else who does this [teaching and presenting short courses for a particular institution]. Then [they] ask me to do it, and then I do it' (Author's translation).

It appears that participants are quite aware that their work is often not of a particularly high standard and falls short of the goal of serving society. Yet they endure in order to serve their own and their employer's hard interests or, stated differently, to meet their goal of earning a living and the various needs and wants met by such remuneration. The following comment is indicative of disillusionment with the work being done. This participant expressed a clear disgust with the 'products' she is supposed to be 'selling': 'You constantly have to sell yourself out there and I don't really know what to sell, to be quite honest' (Author's translation).

Most seem to largely discard softer interests while utilising the accessible though less legitimate means discussed earlier in order to service hard interests. The governing discourses and practices, notions of legislative compliance, science, community-based disaster risk management and so forth, combined with the very nature of contract work, in particular the relationship between time and money, entrench a particular set of instrumentalised practices in which each of these means, in a very narrow and convenient interpretation, become performance indicators in and of themselves. This particular status quo simply has no room for even the most well-intentioned practitioner to address the overwhelming structural complexity that is the danger and processes and afflictions intertwined with danger that many

South Africans experience. Therefore, such well-intentioned practitioners have lacked the means with which to pursue soft interests from the very beginning. Even armed with all manner of potentially appropriate conceptual tools, these local-level complexities informed by a society generally violent in every possible respect are not the purview of consultancies caught up in a cut-throat battle for self-preservation.

Conclusion

The DRAaM practitioner is implicated in the barbarism regulated by our contemporary neoliberal discursive mechanisms. This barbarism takes on the form of manifold post-apartheid dystopias. First, what may be conceived of as neo-*compradorism* entails an emerging group of practitioners earning a living from DRAaM, often without meaningfully contributing in the manner initially intended. This reality is disturbing when read in conjunction with Chapter 3, which highlights the structural violence of the everyday for many South Africans and urgent need for large-scale social change. Consultants and the officials who employ them are essentially compradors of an enduringly violent and illegitimate South African political economy, where conservative discourse in 1994 shifted from pseudo-scientific notions of differences based on race to governmentality partly based on neoliberalism and particular types of professionalism, though essentially affecting similar binaries. Second, many though not all consultants experience *anomie*, emergent from a disjuncture between goals and legitimate means, as they are driven, through the logic of 'time equals money', towards a principle focus on the hard interests of personal and institutional survival, as opposed to the soft interests inspired by a 'service to humanity' ethos.

Note

1 Giddens (1984) refers to ontological security roughly as a situation in which an actor feels assured and certain regarding her position or place in society. The world makes sense, and they have a particular identity and therefore a socially acceptable *modus operandi* within it.

References

Baran, P. 1957. *The Political Economy of Growth*. New York: Monthly Review Press.
De Waal, A. 1997. *Famine Crimes: Politics and the Disaster Relief Industry in Africa*. Oxford/Bloomington: James Currey/Indiana University Press.

Frank, A. G. 1970. The Wealth and Poverty of Nations: Even Heretics Remain Bound by Traditional Thought. *Economic and Political Weekly*, 5(29/31): 1177–1179.

Giddens, A. 1984. *The Constitution of Society: Outline of the Theory of Structuration*. Cambridge: Polity Press.

Merton, R. K. 1938. Social Structure and Anomie. *American Sociological Review*, 3(5): 672–682.

Conclusion

(Un)framing risk as a techno-political instrument

Summary

Similar to Bankoff's (2001) critique of vulnerability as Western discourse, DRAaM in South Africa must be conceived of as a manifestation of the familiar story of 'us' dominating 'them'. Domination is affected through leveraging power and setting the terms by which power is vested in knowledge. 'Us' in this instance are the wielders of DRAaM code, framing the matter of disaster as one for risk assessment and management. Through the de-politicising and re-politicising implications of risk framing, we have inevitably shaped DRAaM to work for us more than it does for them.

Instrumental reason is shaped by manifold and overlapping modes of governmentality located in time and space. Neoliberalism as a macro-discourse has been prevalent since the mid-1970s, emerging in the North but gradually spreading globally. Its general mode of governing potentiates particular instrumental rationalities. These discursive spaces are often guided by the typically modern, though misguided, pursuit of a mythical singular 'the answer'. By celebrating the individual expert in a post-industrial economy, neoliberal governmentality sets the scene for codified theoretical knowledge to be leveraged as a commodity. The code purportedly holds the key to unlocking knowledge and thus holds the fundamental (post-industrial) logic in place, from which modes of instrumental reason may emerge. To be an expert in a given code is to be employed, deploying a power-laden cypher wherever the customer is mandated to consult such expertise.

In the South African case, institutionalisation of DRR as a risk assessment and management expert-driven process has been highly problematic. Following from international conventions and based on the logic of mainstreaming, DRAaM has failed dismally even when measured on its own terms. If one draws on the insights from Chapters 2 and 3, this particular institutionalisation is even more reprehensible, as DRAaM has been institutionalised largely as inaction in a Global South context characterised by the widespread adversities constitutive of structural violence. The actors involved in this particular industry include municipal officials, consultants, academics, students and many hybrids of these categories, forming a policy-reifying self-referential cycle of the ideal types, academia and practice. Academia, focused on DRAaM, is largely morphing towards practice and is equally complicit in the instrumental logics that are largely informed by policy documents. This is true for teaching and for knowledge production. Policy documents are often treated as theory, with 'multidisciplinarity' being associated with a lack of theory and a lack of cogent scholarship. Control over academia through notions of 'relevance' has meant that discourses such as DRAaM have been adopted largely from the North, while South African academia is expected to work within associated parameters. Being governed in this way limits the ability of local academics to offer nuanced interpretations of reality. Many incessant injustices are taken for granted, and thus academics become complicit in conserving an enduringly illegitimate status quo.

The knowledge produced by practitioners has been explained based on a number of actor-constituting micro-discourses stemming from popular but superficial interpretations of legislative compliance. The primary discourse is the notion of expert knowledge. Consultants have commodified key markers of supposed expertise. 'Science' is one such marker, often abused by consultants drawing on readily available data regardless of its intended purpose. Moreover, when assessments are not entirely desktop based, they typically rely on consultative meetings at very high levels of analysis. Technology, especially in the form of GIS, is another marker of expertise. Maps are conspicuously displayed throughout reports and marketing materials. It appears that many DRAs are GIS driven instead of merely being informed by this technology. The result of many of performances of science, including through the use of GIS, is that analyses are based on invalid data. Multidisciplinarity, though often cited as an essential part of DRR and as a marker of expertise, has been more or less lacking. Consulting diverse expertise is not cost effective. Furthermore, broader participation in the form of CBDRM has been instituted only to a limited extent. In many cases, these practices, too, have suffered from the instrumentality associated with consultancy, where means become commodified ends mediated through the competition inherently part of the industry.

The institutionalisation of DRAaM in South Africa has had adverse consequences for both intended beneficiary and practitioner. Many ordinary South Africans continue to suffer a number of generally structurally enforced afflictions, of which disaster remains one component. A lack of legitimate means through which to meet the goal of institutional survival and the industry's official objectives causes a form of distress to practitioners best conceptualised as *anomie*. Experts seek to secure contracts by underselling one another. In addition, they are largely at the behest of a single expert code, offering limited possibilities for meaningful insight into the particularities of diverse contexts. The result is that quests for institutional survival and professional identity often work against the official objectives of disaster reduction and noble intensions of service to humanity.

State building through the embrace of expert knowledge contracted from the private sector, therefore, has in this instance been inadequate. Experts focused on self-preservation have largely not been able to deliver the envisioned public goods. Therefore, in many instances, the state would likely be better served by using resources currently spent on DRAaM for more open-ended engagement to build good-faith relationships between state and society. This type of relationship would be a resource in and of itself and a more substantial tool for understanding the manifold afflictions plaguing South African society. The ensuing dialogue may or may not reveal danger related to disaster as key concerns in a given context.

The professional is political: implications

Professionalisation, hinged upon an all-embracing code, essentially encrypts into public service and development the largely inappropriate and universalising interpretations in the expert's standard operating procedures. This includes the de-politicising force emanating from an indiscriminate appropriation of the idiom of risk assessment and management. The code becomes an instrument of and for reaffirming the professional and, importantly, an instrument by which means become ends and ends become 'irrelevant'. To frame an issue in this particular code is to a significant degree to deny the political nature of social marginalisation and its many synonyms, at least to a significant extent. Danger has been removed from the realm of the politics of deliberation. Instead it is shifted into the realm of technocracy and expert knowledge with all of its vested interests.

However, this burying of politics under bureaucratic procedure is dictated by the terms and practices stemming from international development conventions. Therefore, it is fundamentally political. Not only does DRAaM dismiss any meaningful change at the expense of the enfranchised, including in no small way the often well-meaning expert. It also reconfigures the social realm in favour of these experts and their clients. Power configurations, for the time being, assert the legitimacy of this interpretation of contemporary realities, and thus the expert is legitimately able to impose her constitutive discourses upon others. In the process, a potential opportunity for constructive deliberation, based upon the willingness by all parties to adjust and reduce asymmetrical access to power, is lost. The subordinate are now 'vulnerable', also to an objectifying code. *Difference*, subjected to uneven power relations, is more or less advertently equated to *inferiority*. At the same time, the objectified 'beneficiary' is denied meaningful contribution to definitions of reality. They are, in other words, what we say they are. The only matter to be determined is *how* this pre-ordained conclusion is justified. This is not too much of a challenge, though, for experts in a broadly accepted contemporary discourse. Moreover, the client commissioned these expert insights based upon an intrinsic belief in the code, mandated through policy.

It is here where there remains much scope for future research. The politics of reality and the right to ontology in such an unequal society are central to the future of South Africa. To be sure, much work remains in devising a politics by which this society can be unshackled from its protracted and widespread structural violence. Exactly how does disaster feature in these diverse contexts of precariousness? Insights from such studies may better serve contextualised praxes in aid of the previously mentioned good-faith state–society relationships in a context of structural inequality, continually reproduced through the secular rhythms of the everyday.

There is a danger in dismissing the case study presented here as unique and 'not relevant' to other contexts. Certainly, each context is unique and many of the lessons drawn from one context may not translate to others. I would however argue that lessons from this case study are relevant to other contexts for a number of reasons. First, the politics of ontology also extends to the relationship between the North and the South more generally. Once a noble idea such as disaster reduction is subjected to the vagaries of global governance, there is a real and demonstrated threat that it is distilled, through policy, into a set of generic performance indicators and universal knowledge claims. An archive of knowledge emerges that may be exploited by experts in a global context where expert knowledge is celebrated, and the market influence on the public service and public institutions such as universities are hard to deny. This potentiates asymmetrical politics between an expert with insufficient conceptual resources and in too much of a rush between projects to adequately engage with any particular context. As such, the expert may be set up for failure at the very outset, while much activity unfolds in the name of 'the vulnerable', but with limited tangible results for the intended beneficiaries.

My argument in the case of South Africa has been for structural change, and it is perhaps here where arguments of South Africa's 'exceptionalism' may be voiced most strongly. Johnson (1996), amongst others, however, makes an argument which may cast a different light on the matter. He notes that South Africa is 'a microcosm of the world. There is no other country on earth whose characteristics, and the difficulties they create, are closer to those of the world as a whole'. The characteristics in question relate to race, language and politics and, most importantly, power.

Poverty and inequality and their relationship to race and culture are significant. Today the distribution of resources is more unequal than ever. There is a long history of discourses such as apartheid, the Cold War or 'efficiency' acting as vehicles for sustaining such disparities. If South Africa is indeed a view into the future more broadly, then current and increasingly popular contempt of the status quo, manifested through protest both violent and non-violent, should be a warning (See for example Luckett and Mzobe, 2016). Leaving inequality and structural violence intact, bar cosmetic intervention through technocracy, does not defer crisis indefinitely. Self-interest prevails and, as such, life is reconstituted daily in favour of those with greater access to power. Power has largely coincided with Western modes of knowing and domination in the service of 'civilisation' and resource extraction. Poverty and marginalisation of course manifest differently across the globe, just as they manifest in different though often overlapping ways in South Africa. That there is widespread destitution stemming from the very (dis) functioning of society as opposed to merely poor policy implementation has been demonstrated in the case of South Africa. If one agrees with the likes of Johnson, then this finding may be extended more broadly with similar implications for interventions.

References

Bankoff, G. 2001. Rendering the World Unsafe: 'Vulnerability' as Western Discourse. *Disasters*, 25(1): 19–35.

Johnson, P. 1996. *Modern Times: A History of the World from the 1920s to the 1990s*. London: Phoenix.

Luckett, T. and Mzobe, D. 2016. #OutsourcingMustFall: The Role of Workers in the 2015 Protest Wave at South African Universities. *Global Labour Journal*, 7(1): 94–99.

Index

Note: Page numbers in italic indicate a figure or table on the corresponding page.